Making Complex Machinery Move: automatic programming and motion planning

ROBOTICS AND MECHATRONICS SERIES

Series Editor: **Professor J. Billingsley**
University of Southern Queensland, Australia

1. Making Complex Machinery Move: automatic programming and motion planning
David A. Sanders

Making Complex Machinery Move:

automatic programming and motion planning

David A. Sanders
University of Portsmouth, England

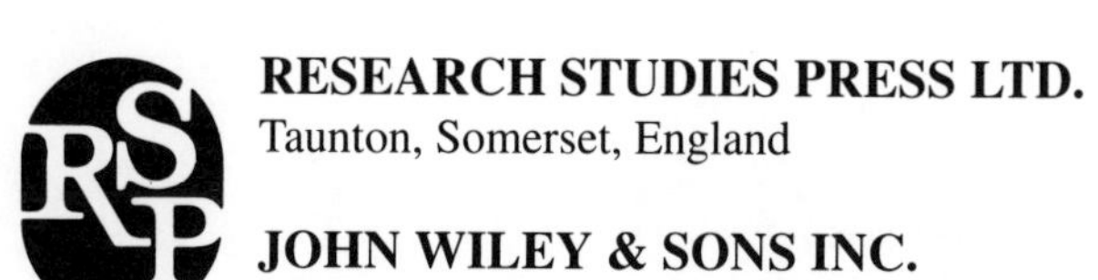

RESEARCH STUDIES PRESS LTD.
Taunton, Somerset, England

JOHN WILEY & SONS INC.
New York · Chichester · Toronto · Brisbane · Singapore

RESEARCH STUDIES PRESS LTD.
24 Belvedere Road, Taunton, Somerset, England TA1 1HD

Marketing and Distribution:

Australia and New Zealand:
Jacaranda Wiley Ltd.
GPO Box 859, Brisbane, Queensland 4001, Australia

Canada:
JOHN WILEY & SONS CANADA LIMITED
22 Worcester Road, Rexdale, Ontario, Canada

Europe, Africa, Middle East and Japan:
JOHN WILEY & SONS LIMITED
Baffins Lane, Chichester, West Sussex, England

North and South America:
JOHN WILEY & SONS INC.
605 Third Avenue, New York, NY 10158, USA

South East Asia:
JOHN WILEY & SONS (SEA) PTE LTD.
37 Jalan Pemimpin 05-04
Block B Union Industrial Building, Singapore 2057

Library of Congress Cataloging-in-Publication Data

Sanders, David A. (David Adrian), 1958–
Making complex machinery move : automatic programming and motion planning / David A. Sanders.
p. cm. — (Robotics and mechatronics series ; 1)
Includes bibliographical references and index.
ISBN 0-471-93793-2 (Wiley)
1. Robots—Motion. 2. Robots—Programming. I. Title.
II. Series.
TJ211.4.S26 1993
629.8'92—dc20 92-26840
CIP

British Library Cataloguing in Publication Data

A catalogue record for this book
is available from the British Library.

ISBN 0 86380 141 2 (Research Studies Press Ltd.)
ISBN 0 471 93793 2 (John Wiley & Sons Inc.)

Printed in Great Britain by SRP Ltd., Exeter

Editorial Foreword

The boundaries are blurring between what were discrete disciplines within engineering. As products gain "intelligence", the designer must create a balance of mechanisms, electronics and algorithms of embedded computer systems, which demands a specialisation all of its own, mechatronics.

Somewhere within the bounds of mechatronics can be found the discipline of robotics. It is debatable whether robotics is characterised by its concern with production machinery (robotnik, the worker) or whether it implies a degree of autonomy. All too often a course in robotics will be dominated by studies of kinematics and their inverse, perhaps because examination questions are easy to define. The fascination arises when sensory systems must be devised and linked to the manipulator — and that is the focus of David Sanders' book.

The vision sensing material is drawn both from research and from undergraduate laboratory experiments which Dr Sanders has set up over the past few years. The equipment is not costly, the techniques are not mystical and the whole is a book with great tutorial value.

John Billingsley
University of Southern Queensland
November 1992

Author's Preface

Brady(1985) defines Robotics as :

The intelligent connection of perception to action.

while a different definition is provided by McKerrow(1990):

Robotics is the discipline which involves:

(a) *the design, manufacture, control and programming of robots;*
(b) *the use of robots to solve problems;*
(c) *the study of the control processes, sensors, and algorithms used in humans, animals and machines; and*
(d) *the application of these control processes and algorithms to the design of robots.*

The study of complex machinery and robots is often referred to as Automation and Robotics. There are several definitions of this subject area (sometimes simply referred to as Robotics). Here we will take the former, broader view, which separates machines from human beings and from animals.

The broad discipline of Automation & Robotics includes aspects of electrical engineering, mathematics, computer science, physics and mechanical engineering. We will attempt to cross the artificial boundaries between some of these aspects. A considerable amount of research has taken place between 1969 and 1991 and a lot of the lower level control problems and the higher planning problems have been solved during this time. We hope to combine some of these solutions and add to our knowledge in the important areas of programming and automatic motion planning for robots and automated machinery. Firstly though ... Why is motion planning necessary?

The time spent on off-line programming, even using computer aided design and simulation packages is expensive. This time could be significantly decreased if the motions were planned automatically and the robots and machinery could be directly programmed from the planning computers. This would also reduce production down time during on-line programming as programs could be downloaded automatically.

In complex and flexible manufacturing environments, tasks may be dynamically reconfigured. In this situation a robot or complex machine may have to stop or may even cause damage. If production is to continue in a flexible environment, the machine controllers need to have their movements planned automatically in order to avoid obstacles and rendezvous with changing target points.

These are the main reasons for research and discussion in these areas. This book presents a novel automatic motion planning system and new path planning and advanced automatic programming methods which take into account both kinematic and dynamic constraints.

The main part of the system comprises a Path Planner and Path Adapter, both using a dynamic World Model. This World Model can be updated by sensor systems, for example a vision system, or from a computer simulation package. The Path Planner contains a geometric model of the static environment and the machinery. Given a task, the Path Planner calculates an efficient collision-free path. This is passed to the control computer where a trajectory is generated and then the machine motions take place.

Predetermination of optimum paths using established techniques frequently involves time penalties which are unacceptably long. Often the original situation has changed by the time a motion is planned. To overcome this problem the automatic path refinement techniques avoid the necessity for optimality before beginning a movement. Repeated improvements to the sub-optimal paths initially generated by the Path Planner are made until the robot is ready to begin the new path. Algorithms are presented which give a rapid solution for simplified obstacle models. The algorithms are robust and are especially suitable for repetitive tasks.

Within the Path Planner, the complex machinery structure is modelled either as connected cylinders and spheres or as connected polyhedral shapes. The range of motion is quantised. The path calculated initially only takes account of geometric, kinematic and obstacle constraints. Although this path is sub-optimal, the calculation time is short. The path avoids obstacles and seeks the "shortest" path in terms of total actuator movement. One of the new path planning methods described in this book employs a local method, taking a "best guess" at a path through a 2-D space for two joints and then calculating a path for the third joint such that obstacles are avoided. Two different approaches are global and depend on searching a 3-D graph of quantised joint space.

The Path Planner works in real time. If there is enough time available before motion begins, a Path Adapter modifies the planned path in an effort to improve the path subject to selected criteria. The Path Adapter considers dynamic constraints and is based on a set of adaptive rules based on simplified dynamic software models of the robot stored within the planning computer. The adapted path is used to automatically program the control computer.

The static model of the robot work-cell is held in computer memory as several solid polyhedra. With the aid of a computer simulation package or a vision system, this model is updated as objects enter or leave the work-place. Overlapping spheres, polyhedral shapes and 2-D slices in joint space are used to model obstacles. In these forms the vision system can be updated and the obstacle data can be accessed efficiently by the path planning and path improvement algorithms.

Together the World Model, the Path Planner and the Path Adapter form an automatic motion planner. Combining the motion planner with inputs from simulators, sensors and vision systems produces a system capable of the most advanced programming techniques for automated machinery and robots.

None of the new systems described here are in use in industry at the time of writing but ... two manufacturers of computer simulation packages are expected to include these methods within two years.

Automatic robot programming should be in place by the year 2000.

This book is dedicated to my parents: John Sanders and Violet May Sanders
and to my sister: Mrs Heather Clothier

Definitions of an INDUSTRIAL ROBOT: There are several definitions of the term industrial robot. In this book the definition of the British Robot Association is used:

The industrial robot is a reprogrammable device designed to both manipulate and transport parts, tools or specialised manufacturing implements through variable programmed motions for the performance of specific manufacturing tasks.

This is stated more briefly by the Department of Trade and Industry as:

A robot is a reprogrammable mechanical manipulator.

Other definitions are :-

The Robotics Institute of America, (now the Robotics Industries Association):

A robot is a reprogrammable multi-function manipulator designed to move materials, parts, tools, or specialized devices, through variable programmed motions for the performance of a variety of tasks.

The marketing division of the Sirius Cybernetics Corporation. **Adams(1979)**

Your Plastic Pal who's Fun to be with!

Acknowledgements

I wish to express my gratitude towards the Automation, Robotics & Manufacturing Research Group at Portsmouth Polytechnic. Specifically to:

Professor John Billingsley *for guidance and inspiration throughout the work which led to this book being published and for introducing me to the joys of wind surfing.* **Dr Bing Luk** *for ALWAYS being available with helpful advice, an optimistic view and to assist in eating any food.* **Mr and Mrs Ali Husseinmardi** *for help and advice on the more practical aspects of the early work and for introducing me to behtarin mehmannavazie Irani.* **Dr Dave Robinson** *for his constructive advice and for introducing me to the extra curricular items available at research conferences abroad.*

Also to

Mr Tom Cronk,	**Mr Ron Dadd,**	**Dr David Harrison,**
Mr Mark Jaques,	**Mr Kevin West,**	**Mr Barry Haynes,**
Mr Antony Moore,	**Mr Ian Stott,**	**Mr Abraham Mazharsolook,**
Mr Peter Harris,	**Mr Tim White,**	**Mr Giles Tewkesbury,**
Dr Fazel Naghdy,	**Mr Niel Bevan,**	**Mr Paul Strickland,**
Dr Jim Hollis,	**Mr KP Liu,**	**Mr Arthur Collie.**

for creating a good working environment and for putting up with me!

and to **Ms Karen Covell** *for proof reading most of this.*

Contents

1. **Introduction.**

1.1	**Automatic Motion Planning: The Requirement.**	*1*
1.2	**Automatic Motion Planning: A Description.**	*4*
1.3	**Automatic Motion Planning: Some New Developments.**	*8*
1.4	**The Path: Obstacle Detection.**	*11*
1.5	**The Path: Planning in Joint or Cartesian Space.**	*15*
1.6	**The Path: Criteria for Improvement.**	*18*

2. **Background.**

2.1	**Introduction.**	*20*
2.2	**Modelling: Obstacles and the Static Environment.**	*21*
2.3	**Modelling: The Robot.**	*24*
2.4	**Path Planning.**	*25*
2.5	**Path Optimisation.**	*33*

3. **Development of a Typical System.**

3.1	**Introduction.**	*40*
3.2	**The Development of an Apparatus: A Servo Amplifier**	*42*
3.3	**The Development of an Apparatus: Computers and a Robot**	*48*
3.4	**The Software Systems.**	*49*
3.5	**Communications.**	*56*
3.6	**The Inclusion of a Bus System.**	*57*
3.7	**An Example Robot.**	*59*
3.8	**The Joint Servo Controller.**	*66*

4. **Modelling of the Robot and Obstacles.**

4.1	**Introduction.**	*69*
4.2	**The Static Environment.**	*70*
4.3	**The Robot.**	*71*
4.4	**Dynamic Obstacles.**	*73*
4.5	**The Transformation into Joint Space: Introduction.**	*75*
4.6	**The Transformation into Joint Space: Spheres.**	*77*
4.7	**The Transformation into Joint Space: 2-D Slices.**	*81*
4.8	**The Transformation into Joint Space: Other Models.**	*83*
4.9	**Results.**	*87*
4.10	**Discussion and Conclusions.**	*100*

5. **Sensor Fusion and Force Sensing.**

5.1	**Introduction.**	*109*
5.2	**Sensor Fusion.**	*111*
5.3	**A Test Rig to Investigate Sensor Fusion.**	*113*
5.4	**A Method of Fusing Sensor Information.**	*116*
5.5	**Typical Conflicts in the Data from Sensors.**	*118*
5.6	**Motion Improvement Using Force Information.**	*120*
5.7	**Obstacle Detection Using Force Sensing.**	*129*

6. **Image Data Processing and A Vision System.**

6.1	**Introduction.**	*135*
6.2	**Overview of an Apparatus (Including a Vision System).**	*136*
6.3	**Obstacle Detection: The Configuration of the Apparatus.**	*138*
6.4	**Obstacle Detection: Low Level Vision Techniques.**	*145*
6.5	**Obstacle Detection: High Level Vision Techniques.**	*151*
6.6	**Results.**	*158*
6.7	**Discussion and Conclusions.**	*162*

7. **Path Planning.**

7.1 **Introduction.** *164*

7.2 **Path Planning for a Single Link Manipulator.** *167*

7.3 **Extension to 3-SPACE Local Heuristic Methods.** *177*

7.4 **Extension to a 3-SPACE Global Method.** *178*

7.5 **Trajectory Generation.** *185*

7.6 **Results.** *186*

7.7 **Discussion and Conclusions.** *190*

8 **Path Improvement Considering the Manipulator Dynamics.**

8.1 **Introduction.** *194*

8.2 **The Lagrangian Formulation for a Manipulator with Three Joints.** *197*

8.3 **The Formulation of Experiments to Determine the Dynamic Model.** *203*

8.4 **An Experimental Method to Determine the Dynamic Model.** *206*

8.5 **Results.** *209*

8.6 **Application of the Model to Improve the Motions.** *219*

8.7 **Results from the Application of the Simple Rules.** *222*

8.8 **Discussion and Conclusions.** *224*

9. **Discussion and a Look to the Future.**

9.1 **Introduction.** *226*

9.2 **Discussion and Review.** *227*

(a) **Modelling of a Robot and Obstacles.** *227*

(b) **Image Data Processing and The Vision System.** *228*

(c) **The Systems and the Apparatus** *229*

(d) **Automatic Path Planning.** *230*

(e) **Sensor Fusion.** *231*

(f) **Force Sensing.** *234*

(g) **Path Improvement to Minimise Peaks in Joint Motor Currents.** *236*

(h) **Path Improvement Considering the Robot Dynamic Equations.** *236*

9.3 **A Look to the Future.** *237*

Appendices

A: The Detail of the Transformation Programs. *242*

B: Edge Following and Line Fitting. *246*

C: The Mechanical Detail of the Manipulator. *248*

D: Specifications of the Mitsubishi RM 501 Robot. *255*

References. *259*

Index. *266*

Chapter One

INTRODUCTION

1.1 Automatic Motion Planning: The Requirement.

Reprogrammable machinery and Industrial Robots are generally more costly than dedicated handling machinery for pick and place and assembly operations. Their claim to superiority is through their ability to be reprogrammed to carry out a variety of tasks, but when complex automated machinery and robots are used, they tend only to be programmed for a finite repertoire of tasks. They have little autonomy. Although the technology is suitable for many areas of industry, including Flexible Manufacturing Systems (FMS), the costs at present associated with installation and the perceived programming complexity exclude them from many applications [Policy Studies Institute(1986)].

We will describe improvements to automatic programming, motion planning and control methods. These methods will decrease the complexity of installation and reprogramming while reducing the associated set-up and running costs. The methods will allow manufacturing and industrial systems to deal with unexpected situations in unstructured environments. These improvements will help to justify their use in smaller factories and for new and wider applications including small batch manufacturing.

The applications for robots and complicated machines are becoming increasingly more elaborate. Industrial pressures are to use the technology efficiently and to reduce process down-time. At present it is often necessary to stop the machinery during reprogramming. Path planning and programming are carried out by human operators or by human programmers on off-line systems. Little motion planning takes place once a path has been selected. Human programmers construct programs to carry out a task in one of three ways:

(a) **Lead through**: The robot is directed along the path by a human operator holding the end effector, and a computer records the joint coordinates at discrete intervals or at specific configurations. This method could be used for an application such as paint spraying, where some of the knowledge and skills of the human operator are required.

(b) **Teach pendant**: The robot is directed along the path by a human operator using manual controls, usually switches. The coordinates are recorded at discrete intervals. This method could be used for an application such as spot welding, where the points for the spot weld are important but the path between the points may not be critical.

(c) **Off-line programming**: The path is defined from a computer simulation of the robot and the work-place. The trajectory locus is then down-loaded to the machinery controller. This method could be used for parts handling and transfer, where the relative positions of objects and the path of the robot or machine are important to avoid collisions.

Using these methods to develop new robot programs can be expensive, time-consuming and tedious. Programming and reprogramming form a large part of the total cost associated with establishing a robot cell, and this cost must be less than the cost of completing the task by other means. When robots are used for repetitive jobs, the costs are spread over many operations and this has contributed to limiting the use of robots to mainly repetitive tasks.

In all three methods of programming, including off-line programming, (even using CAD packages with interactive graphics), it is the responsibility of the operator to choose the via points so that a robot path both avoids collisions and is efficient.

In the future these programming methods may not be satisfactory. Time spent reprogramming will be expensive. Complex robot systems as part of FMS may have to adapt to new tasks in the work-place on-line and without human intervention. Other applications make it difficult for a human programmer to intervene in, for example, work in undersea, nuclear and space environments. More complex applications will require robots and machines to work around each other while avoiding collisions.

Most robot applications still involve repetitive tasks, and Udupa(1977) introduced the infrequent initialisation hypothesis. This hypothesis suggests that, in general, a large number of robot trajectories will tend to be planned in any given environment before the environment changes. This suggests robot paths are especially suited to on-line reprogramming and the automatic and adaptive motion planning procedures we describe.

As part of his research, Kumar(1988) completed a short study of robot programming requirements and described the following desirable characteristics for a system. The system must:

(a) *be capable of successful and precise execution of a specified task.*

(b) *be versatile and able to adapt to different tasks, as well as to a changing environment.*

(c) *execute the tasks in the most efficient manner, where the definition of efficiency could be flexible.*

This book explains new methods of automatically programming robots and complex automated machinery; the methods have the advantages of eliminating programming cost for new paths, reducing down-time and set-up time and allowing robots to be used for changing tasks in changing environments. The machinery motions are automatically reprogrammed between tasks and whenever the environment changes. These paths can then be adapted to improve the motion.

The result can be expected to be safer and more efficient in comparison with the other programming methods discussed.

The ability automatically to plan collision-free motions for a robot manipulator is one of the capabilities required to achieve task-level robot programming. Task-level programming is one of the principal goals of robotics research. It is the ability to specify the robot motions required to achieve a task in terms of task-level commands, such as:-

"Move BOX to TABLE"

rather than robot-level commands, such as:-

"GOTO 2.5, 6.3, 41.7, 36, 42, 90".

1.2 Automatic Motion Planning: A Description.

The motion planning problem, in its simplest form, is to find a path from a specified START configuration to a specified GOAL configuration that avoids collisions with obstacles, and then to generate a set of trajectory loci to guide the machinery actuators. This problem is more complicated than the collision detection problem where a known robot configuration is tested for an impact. Automatic motion planning is also dissimilar from on-line obstacle avoidance which entails revising a known motion so as to circumvent unforeseen obstacles. Both these problems have now been solved for a number of particular machines and robots, and systems are in use in industry. Automatic motion planning is only just emerging from the research stage.

Motion planning depends on the subjects of "Path Planning" and "Trajectory Generation and Tracking". Research in the USA and Europe into these areas has developed along two parallel lines in isolation from each other. The work and the new ideas described in this book attempt to cross this artificial and unnecessary divide.

Generally a robot path planner performs higher-level functions, breaking the task into a sequence of smaller movements based on its knowledge of the environment and the capabilities of the manipulator. The inputs to the planner are a description of the state of the manipulator and the environment. Based on these inputs, it plans the task. The controller then moves the machinery through the planned configurations.

Since processor speeds are limited, the planners have not as yet been expected to output a trajectory locus at a speed which the controller can follow. A more practical arrangement has been for the motion planner to work off-line and refuse new information during execution. All relevant information must be provided at the beginning of the planning program and the planner plans the path for the whole task. The planner works in near real time. The path is improved using information on robot dynamics that has usually only been used in the past for the optimisation of robot trajectories.

The trend in robotics research has been to achieve the "Task" by partitioning the motion planning problem into three stages, the input to each stage being the output of its predecessor. The three stages may be considered as shown below:

Path Planning

Given a Task for a robot and a geometric description of its environment, plan a path that avoids collision with obstacles. The path is a function of space.

|

Trajectory Planning

Given a path to be followed by the end-effector, the actuator constraints and a dynamic description of the robot, find the positions and velocities of the joints to achieve the path. The trajectory locus describes trajectory curve in joint space and the trajectory specifies the robot configuration as a function of time and space.

|

Controller Trajectory Tracking

Given a trajectory and the dynamics of the robot manipulator, track the given trajectory by servoing the movement of physical actuators.

This suggests that future work-cells within fully automated manufacturing systems might be as shown in figure 1.1. These work-cells will require automatic programming and reprogramming systems.

Many papers on path planning have appeared in the literature during the 1980s. While much work has been done, progress towards efficient general techniques has been slow. This lack of progress appears to be due to the complexity of the problem. In computer complexity theory Hopcroft et al(1984(b)), Shwartz & Sharir(1984) and Shwartz & Yap(1987) have stated that the problem is exponential in the number of degrees of freedom.

The path planning problem has generally been regarded as a purely geometric problem and has tended to be considered by computer scientists. Path planning has involved building a geometric model of the world and a free-moving object. These models are then used by procedures for determining spatial paths across the World Model. Any system dynamics are ignored by this approach and few researchers have considered complex multi-link machines. This book will!

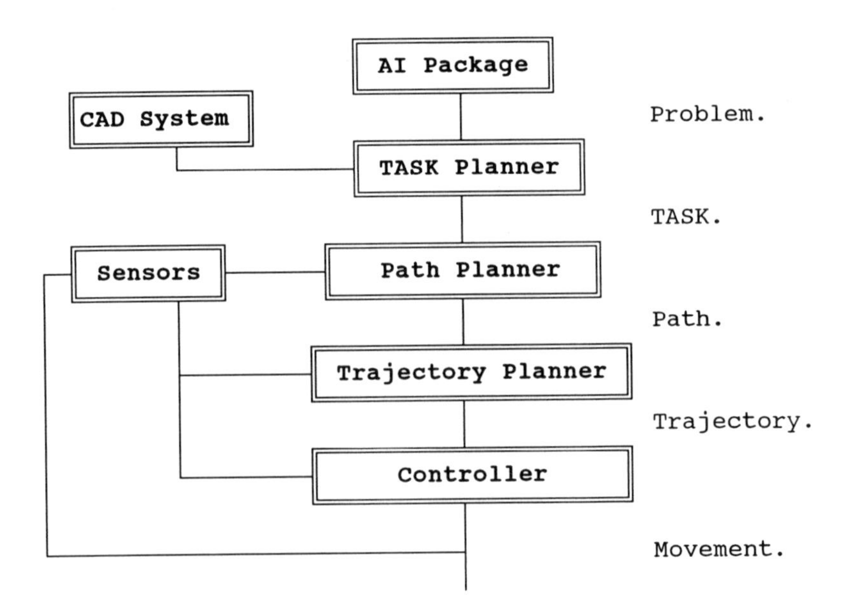

Figure 1.1: A Work-Cell within a Fully Automated Manufacturing System.

The trajectory planning problem has been considered by control engineers and assumes the path has already been planned. Trajectory planning has been concerned with the manipulator dynamics, not with geometric world models.

Considerable advances have been made in path planning and trajectory planning, yet there have been few attempts to combine the two. In recent years the need for combining them has been recognised, [Schwartz & Yap(1987), Brady et al(1989)] but processor speeds and economic factors still dictate that the planning stage be isolated from the tracking stage, [Brady et al(1982), Kumar(1988)].

Reducing the path planning problem to pure geometry allows a very precise problem statement and solution, but important non-geometric constraints are not considered. Solving the geometric problem is computationally fast, but in simple static environments, path optimisation may be more important than planning time. In a complex industrial FMS solution speed may be more important than path optimality. The methods described later satisfy both these situations.

It is accepted that economic factors and processing speeds still remove the path planning problem from the controller and on-line optimal planning has not been possible. Instead, we present automatic and adaptive reprogramming methods which consider dynamic constraints as well as geometric and obstacle constraints, but which do not attempt to optimise the path before beginning any movement. A path is found which is only improved if time is available before motion begins.

In most published work dealing with off-line motion planning techniques, the calculation time is not critical. This book describes an automatic system which will work in real time and allow the advantages of a truly flexible manufacturing system to be realised. In a highly flexible system, any changes require fast re-calculation of machinery motions and robot paths. The extent to which this can be achieved may determine the re-scheduling capabilities of an entire FMS or factory.

In order to achieve real-time operation a compromise is initially made between the efficiency of the calculated path and the calculation time. For any motion planning problem there is an optimum solution based on a chosen cost function. In a changing situation operational constraints make a faster sub-optimal solution more acceptable. Initially a fast sub-optimal path is produced which is only improved if time allows. The sub-optimal solutions ensure that the calculated path is collision-free and tends to the shortest path in terms of total joint movement.

The work described can also be used within off-line path planners to reduce the reprogramming time and produce more efficient paths than the human operator.

1.3 Automatic Motion Planning: Some New Developments.

Some research has been devoted to automatic motion planning over the last five years, but few of the methods are simple enough and powerful enough to be practical. Algorithms are particularly scarce for machines which form kinematic chains with revolute joints, (the most popular type of industrial robot).

With the exception of Khatib(1986) the motion planning work completed in the past has required computation time that makes the robot wait before carrying out the planned trajectories. The methods presented allow the robot to continue working, and new movements are automatically planned and improved as necessary.

Several methods of automatic path planning are described to provide a comparison of the two main classes of path finding algorithm: local, heuristic methods and global methods. The problems experienced by Khatib using his artificial repulsion approach are overcome and some of the initial work is similar to work presented by Balding(1987) in that specific configurations are represented by nodes within the local methods, and the configuration space consists of a lattice of points. The space is discretised within the global method and each node in the lattice represents a small neighbourhood in configuration space. The total space will be called a Configuration Space Graph (CSG).

The local and heuristic path planning procedures produce real-time solutions for a range of problems. They require little time to pre-process data to generate a 2-SPACE graph before searching for a path. The methods can be employed in circumstances where the environment changes frequently or the environment is unknown and information is being received from sensor systems.

The global path planning methods are established on a more rigorous mathematical treatment of the path finding problem. The methods provide solutions which consider only the constraints of the obstacles and the restrictions of the world model. They require time to pre-process data to generate the world model as a 3-SPACE graph in joint space but then furnish real-time answers to a range of motion planning problems.

Once a sub-optimal path has been planned quickly by one of the path planning methods, the path is improved by considering other constraints of the robot manipulator. Other constraints considered in the original work were the maximum actuator velocities and accelerations and the machinery dynamics.

The control computer initially receives input from the fast path planner and then from a slower but more efficient path adapter. The whole motion planning system accounts for both obstacles and dynamic constraints, and produces control signals for the machinery actuators.

The remainder of this introduction discusses the preliminaries to the three major areas of automatic motion planning considered. This will aid in understanding the original work and the background and literature survey presented in Chapter two. The preliminaries are obstacle detection methods, the choice of space for path planning and criteria for path optimisation. Chapter two is an extensive literature survey of the relevant previous work and it provides a background for the work described in the rest of the book.

In Chapter three the development of the hardware and the systems is considered from an initial test rig to the final apparatus. As part of the work a novel parallel hierarchy control structure was developed and this is described. In Chapter four the decisions on the types of model to use for the robot, obstacles and the environment are dealt with. In this chapter the new concept of using diverse models for different parts of the work-place is introduced. The models for the static environment were complex and time-consuming while the changing environment could be modelled in a fast, simple and novel way. Because the off-line programmers available in industry tend to use polyhedral models, these are also discussed. In Chapter five, sensor systems are discussed, concentrating on how to interpret the data and to resolve conflicts between sensors. Methods of detecting the joint motor currents are also considered and the differences between transients caused by collision and those caused by changes in direction are described. This was used as a final check that all was well and a collision was not taking place.

The vision system and methods of 3-D visual data processing and image acquisition are described in Chapter six, using the models selected from Chapter four. Techniques are developed for incorporating the obstacle detection data into the decision-making process of the path planner.

Chapter seven presents the novel methods of automatic path planning. Several multi-degree of freedom path planning algorithms are described. These are developed from initial work considering two-dimensional graphs of joint space. Several 3-SPACE methods presented are local and heuristic methods, and two are global methods. One global method uses the polyhedral models commonly found within industrial simulation systems and the other uses the novel models discussed in Chapter four.

Identification of the parameters of the manipulator dynamics is studied in Chapter eight. These parameters are used to influence the strategy of the path adapter. The identification procedure is demonstrated experimentally on a Mitsubishi RM-501 robot. This method depends on simplified dynamics models of the robot which are used to develop simple adaption rules. These rules are stored in the main computer and used to adapt the path in order to reduce the time taken to achieve the task. In the work described in the literature the dynamics of a manipulator have only been used to adapt a trajectory produced from some planned path. In this work, the dynamics are used at a higher level, within the path planner.

In Chapter nine the work is reviewed and discussed. Conclusions are presented, along with a description of work planned for the future. The algorithms introduced have a number of advantages: they are simple to implement, they are fast for machinery with few degrees of freedom, they can deal with machinery having many degrees of freedom (including redundancy), and they can deal with confused, unknown and changing conditions. Finally, it is demonstrated that the whole system can be employed for real-time control.

1.4 The Path: Obstacle Detection.

The work-space of a manipulator includes all possible physical elements swept by the links of the machinery as the actuator (position) values vary from their minimum to maximum values. This work-space can include static and dynamic obstacles. Static work-space environments include such things as the floor and walls. These do not change with time and may be modelled by complex and accurate methods. Dynamic obstacles change with time and even simple fixed sequence machines and robots may require sophisticated obstacle detection and avoidance techniques to deal with them.

In automatic motion planning, a prerequisite to circumventing any obstacle is to detect it. Various methods for detecting obstacles have been proposed by different authors including Udupa(1977) and Doty & Govindaraj(1982). These methods are now in use and are described generally in several text books, including: Fu, Gonzalez & Lee(1987), Klafter, Chmielewski & Negin(1989), Galbiati(1990) and Mckerrow(1990).

Several detection methods are considered. These are detection by:

(a) Human Operator.
(b) Ranging and proximity detectors.
(c) Force feedback.
(d) Vision Systems.

These detection methods are considered in this Section, followed by some brief discussion and conclusions.

(a) Detection by a human operator. In this case the operator determines the obstacles in the work-space that may interfere with the manipulator. If the system is under direct operator control, this information is recognised by the human operator and revised commands are sent to the robot.

If the trajectories are computer generated and computer controlled, then the information about the obstacles has to be manually loaded into the computer to create a World Model. During the planning of trajectories the computer checks against this stored information for possible collisions. This method is suitable only for environments which rarely change. Whenever the work-space environment changes, the system depends on the operator to update the information about the new work-space environment. The accuracy of the World Model depends solely on the operator and involves tedious and complicated surveying and entry of information. These data are often corrupted by human error.

(b) Detection using ranging and proximity detectors. These detectors involve various technologies: light, acoustic, infra-red, etc. All ranging systems employ a transceiver to update the world model. Some methods are described in Klafter, Chmielewski & Negin(1989) and McKerrow(1990).

Obstacle detection ranging systems take evasive action if the robot comes within the minimum safe distance from an obstacle. This method can be used for monitoring both static and dynamic environments.

Although the positional accuracy of these devices can be excellent, directivity of the transceivers creates blind spots and limits the detection volume.

(c) Detection by force feedback. If unchecked, the robot will try to overcome any obstacles in its path. In doing so, the various torques - both in the joints of the manipulator and the actuators of the various joints - will increase rapidly. These forces can be detected by using force detectors such as strain gauges placed at the joints [Raibert & Craig(1981)] or by monitoring the actuator torques [Doty & Govindaraj(1982), Sanders(1987(a)), Luk et al(1988), Sanders(1990)]. On detection of abnormal changes, corrective action can be taken.

If the obstacle is movable, it will be forced to move and the torques in the manipulator momentarily increase to overcome the friction and inertia of the object. If the obstacle cannot be moved, the actuators will be torqued more and more until either the obstacle or the manipulator is damaged. Both these results are undesirable. There will be a momentary increase in the torques of the links that are moving at that instant. These torques exceed the torques that will be encountered during normal full load working conditions and evasive action can be taken.

(d) Detection by vision. Vision systems comprise one or more cameras, a controller and, often, special lighting equipment. Cameras are usually placed above the manipulator work-space. Typically the manipulator work-space is brightly lit and often back-lit. The cameras constantly scan the work-space and pass the information to a computer where the information from other sensors may also be correlated. A snapshot of the work-space is formed at discrete time intervals and this is described by Fairhurst(1988), Davies(1990) and Galbiati(1990).

Since a vision system sees both the manipulator and the obstacles it is possible to use a feedback loop from the cameras to control the trajectory of the manipulator. Whenever the machinery comes within the minimum clearance distance of an obstacle, corrective action can be taken. This needs a large amount of fast processing power, but there is no need to store information about the obstacles in large databases.

The main disadvantages of vision systems are their cost and complexity.

Discussion: Considering each of the four methods: Intervention by human operator after obstacle detection will be inefficient for complex, modern and future robot applications working in dynamic environments. However, the initial data for the accurate static World Model can be entered by a human operator after detailed surveying and/or measurement.

The accuracy of detection by ranging or proximity sensors depends on the number, configuration and accuracy of the transducers. To cover a whole work-place a great many ranging devices would be required. This can create further problems when information from different sensors conflicts, and this will be discussed later.

Detection by force sensing is cheap but is only suitable for low cost and very slow moving manipulators or as a "last resort" back-up protection mechanism. There is little application to the path planning problem, but the similar torques experienced as actuators change direction may be used within a path improvement algorithm.

Vision systems tend to be largely independent of the operator and there are several advantages to this method. Vision systems can monitor the machinery and obstacles in a common universal frame. The system constantly keeps track of the work space environment of the manipulator and hence it can be used to monitor dynamic or frequently varying environments. The information can be used directly for collision detection or for updating the path planning computer.

Conclusions: A human operator is the most accurate source of data concerning the detailed static environment. This method was selected to enter the accurate data for the static work-place where time constraints are less important. A vision system was selected to update the dynamic obstacles in the World Model as this detection method is global, covering the whole robot work-space area, while still being fast enough to work in real time. The disadvantages of methods (b) and (c) excluded their use for global obstacle detection, but they are both considered in later chapters.

1.5 The Path: Planning in Joint or Cartesian Space.

Path planning problems are difficult partly because manipulators and obstacles are best described in different spaces. Path planning can take place in either space. Obstacles tend to be described in 3-D Cartesian space but complex automated machinery and robots tend to be especially time-consuming to describe in this space.

A typical manipulator is composed of a number of joints, and the movement from one configuration to another is accomplished by moving each joint. The state of a manipulator may generally be defined as a vector specifying the various joint angles, $\underline{\Theta}$. This representation is called a configuration and is a representation in joint space. For the robot described:

$$\underline{\Theta} = [\Theta_1, \Theta_2, \Theta_3, \Theta_4, \Theta_5]^T.$$

If configurations are specified as a path in Cartesian Space, that is $[X,Y,Z,\Theta,\epsilon,\phi]$, then the controller must compute the required joint configurations through inverse kinematics. If the path planner provides the reference path to the controller in real time, the controller must calculate the velocity transformations from Cartesian space to joint space using complicated algorithms, usually using the inverse Jacobian. The computation of the inverse Jacobian is a non-trivial problem, particularly if it must be done at the frequency at which the controller must operate.

There are several instances where the path is specified in the Cartesian coordinates $[X,Y,Z,\Theta,\epsilon,\phi]$, such as welding or complex parts handling. In such cases, the path must be transformed to joint coordinates $[\Theta_1,\Theta_2,\Theta_3,\Theta_4,\Theta_5]^T$. The Jacobian is only a point transformation and there is no known functional transformation to map the entire path from task space to joint space. This means that the transformation must be accomplished at a certain number of discrete points. Taylor(1976) also investigated this problem at Stanford.

A few controllers have been designed to operate at the task level, but these were slow, and controlling at the joint level is simpler and tends to be faster. The actuator constraints and description of the robot dynamics are in terms of joint coordinates in joint space.

Several authors have made a once-only transformation of the manipulator and its surroundings into some abstract space. Udupa(1977) enlarged obstacles by the width of the manipulator links to produce a primary map. A transformation was applied which permitted the upper arm to be viewed as a point. The transformed space was called a primary chart and was a map of all the positions of the end of the upper arm for which the upper arm was collision-free. The secondary map was produced by enlarging the obstacles by the radius of the forearm.

The advantage of these transformations was that the path planning of a point or single line segment was much easier in these transformed spaces. These methods are used in one of the two global path planning methods described later.

Lozano-Perez & Wesly(1979) and Lozano-Perez(1983) developed a method for the calculation of paths for polyhedral objects moving through a space littered with other polyhedral objects. The method involved transforming obstacles into an abstract space which he called Cspace. An example of how this method is used may be found in Red(1984). In that work the configuration space for a PUMA robot was calculated by a VAX minicomputer. The configuration space was displayed graphically and the operator could plan a path for a point through this space. The path was then converted back into robot coordinates for execution of the task. The method worked off-line.

The configuration of a three-dimensional object may be specified by a six-dimensional vector. The six-dimensional space of configurations for an object **Obj_A** is denoted by **Cspace_A**. This contains all the information necessary to solve the find-path problem for **Obj_A**. Lozano-Perez reported that when an object was a three-dimensional solid which was allowed to rotate, then a simple object **Obj_A** in real space became a complicated curved object in six-dimensional C-space. So he did not calculate such objects; instead he approximated objects by a series of two-dimensional slices containing polyhedral shapes.

Brooks(1983(b)) represented two-dimensional (2-D) free space as a union of possibly overlapping generalised cones. The algorithm translated a polygonal moving body along the axes or spines of the generalised cones and rotated it at the intersections of the generalised cones. The algorithm was fast and generated paths that gave good clearance from obstacles. Brooks, together with Kuan, later improved the quality of the paths found by representing the 2-D free space as a union of generalised cones and convex polygons.

Brooks(1983(a)) transformed the space between the obstacles into freeways for the upper arm and payload of a robot. The two freeway spaces were searched concurrently with the constraint that the upper arm and payload were a fixed distance apart, due to the forearm. Brooks reduced the degree of freedom of the payload in order to simplify the problem. The algorithm generated prisms of free space between obstacles. The obstacles were effectively only two-and-a-half dimensional in that they had a two-dimensional shape and a height. Thus a cube could be represented accurately but a tetrahedron could not.

Conclusion: The purpose of creating a different space through a transformation is to reduce the complexity of the path finding problem. Even with reduced complexity, none of the systems mentioned in this section achieved real-time operation.

In joint space a configuration is represented by a point and the problem is reduced to finding a path for a set of connected single points through a set of obstacles. The limits of the joints form the boundaries of the joint space. Path planning in this space is reduced to finding a collision-free path for a point, and this is a relatively simple problem.

There is little advantage to planning the robot path in Cartesian space, and many advantages to planning paths in joint space. The new work described uses joint space to plan the robot paths and transform obstacles to this space. The methods are described in Chapters four and seven. Other work is described in Lozano-Pérez(1983) and Faverjon(1984).

1.6 The Path: Criteria for Improvement.

Given a task, the objective is to complete it efficiently. For path planning problems investigated by computer scientists, this has usually been interpreted to be by the shortest path. This has been dealt with in a few papers [Lozano-Perez(1981), Sharir & Schorr(1984), Papadimitriou(1985), Bajaj & Moh(1988)], but these researchers tended to be concerned with free-moving objects and not connected chains.

Industrial machines and robots move to a desired goal by moving each joint individually. Any concept of shortest distance, which would be relevant if the machine were a free-moving disc or a sphere, is meaningless. For a robot, any definition of shortest distance may not be the best path in terms of safety or in terms of some dynamic performance such as minimum time or energy. It is also possible that the shortest path may not be achievable.

The torque from the actuators must be considered for path improvement. In classical optimal control theory, the controller is designed in feedback form and, typically, the dynamics of the robot can be expressed in the form

$$\tau = D(d^2\theta/dt^2)(\theta_{(t)}) + h(d\theta/dt,\theta) + c(\theta)$$

where

τ	=	Vector of actuator torques.
θ	=	Vector of joint positions.
$D(d^2\theta/dt^2)$	=	Inertia, acceleration-related symmetric matrix.
$h(d\theta/dt,\theta)$	=	Nonlinear Coriolis and centrifugal force vector.
$c(\theta)$	=	Gravity loading force vector.

This expresses the nonlinear and coupled nature of the differential equations that describe the system. There are also other position, velocity and acceleration dependent constraints imposed on the system. This complexity means the path adapter cannot operate on-line at the speed of the robot control computer.

The actuator torques can also be read directly from currents in the servo-amplifiers connected to the actuators. [Sanders et al(1987(c)) and Sanders et al(1991(a))].

Discussion: In this book the path adapter operates on a planned path to improve the robot performance in terms of some dynamic criterion, for example time. No attempt is made to solve all the dynamic equations in real time. Instead simplified dynamic models of a robot are used to establish simple rules for the path adapter. This work is presented in Chapter eight.

Chapter Two

BACKGROUND

2.1 Introduction.

Automatic robot programming and automatic motion planning algorithms must coordinate the essential aspects of the problem; that is detecting obstacles, modelling the machinery, robots and obstacles and deriving suitable trajectories for the actuators.

Detecting obstacles was considered in chapter one and a vision system is described in some detail in chapter five. This chapter will consider modelling methods and the derivation of suitable trajectories.

Pieper investigated automatic programming for robots in the United States during the late sixties. This work was extended by Udupa at CalTech and Widdoes at Stanford during the seventies. This early research aimed to design a robot programmer for use in planetary exploration. The major contributions in the field of path planning during the eighties have come from Lozano-Perez at MIT and then IBM and from Brooks. Both used polyhedral models to represent obstacles and their original work has been extended more recently by various scientists.

An alternative approach was reported by de Pennington where the mover was modelled by a series of interconnected spheres. Since then other authors have added to this research.

The main parts of path planning research systems have been a world model, the path planning algorithms and the output. The few research systems considering path planning algorithms for robots have tended to use three inputs:

(a) A geometric and kinematic description of a robot.

(b) A geometric description of the robot environment.

(c) The task description.

The type of world model chosen to describe the robot environment has a considerable effect on the path planning algorithms; and different types of models are discussed in Sections **2.2** (Modelling of the Obstacles and Environment) and **2.3** (The Robot Model).

Several different path planning methods have been proposed and these are discussed in Section **2.4**. Finally, previous work in path optimisation is discussed in Section **2.5**.

Sections **2.2** and **2.3** provide the background to the original work presented in chapters four and five of the book, and sections **2.4** and **2.5** provide the background to the original work presented in chapters six, seven and eight.

2.2 Modelling: Obstacles and the Static Environment.

Many computer models are possible. In the fields of computer science, artificial intelligence and robotics research the most popular method of representing objects has been by using "polyhedra", and many examples are included in the references at the end of this book. A polyhedron is a three-dimensional solid figure with many planar faces. The edges where faces meet are linear. Most objects may be closely approximated by polyhedra, and examples of programs which model moving objects and their environments by polyhedra can be seen in the work by Lozano-Perez and in the robot simulation system, GRASP.

A polyhedron may be represented by a tree structure of edges, faces and vertices. An edge may be defined by its end points and a face may be defined by specifying its edges. The more complex the polyhedron the more edges, vertices and faces it has and hence the more data required to define it.

Having detected obstacles in the robot environment, a prerequisite to robot path planning is interference detection. Some original work on interference detection among polyhedral solids was presented by Boyse during the seventies. In this work, to determine whether a polyhedron **Poly_A** intersected a polyhedron **Poly_B**, all the edges of **Poly_A** were tested to see if they intersected any of the sides of **Poly_B**. If **Poly_A** and **Poly_B** were simple cubes then each of the twelve edges of **Poly_A** had to be tested with each of the six faces of **Poly_B**. This gives a total of seventy two edge face tests. A test also had to be done to see if **Poly_A** was enclosed by **Poly_B** or vice versa.

"Solid modelling" has been used to represent the robot work-space, and De Pennington used Constructive Solid Geometry (**CSG**). CSG models use simple shapes, called primitives, to produce complex and accurate representations of a robot's surroundings. The primitives fulfil particular mathematical properties, so that operations such as volume calculations and intersection checking can be carried out easily. An example is shown in chapter four.

Spatial occupancy enumeration (**SOE**) is another subset of solid modelling. Space is divided into a matrix of spatial cells. Each cell is defined either as containing an obstacle or free space. During the eighties, both Ahuja and Dupont have shown that this method can be used to represent the path planning problem. In these works a tree structure was used to represent three-dimensional space. Space was represented as a solid cubic block. This was subdivided into eight blocks. Each block was tested and given a "colour flag". A block was designated black if it was completely within an object, white if it was free space and grey if it contained object and space. Each grey block was then subdivided into another eight blocks. Recursive subdivision continued until a minimum sized block was reached. At this point any minimum sized grey blocks were designated as black.

To solve the collision detection problem using **SOE** the obstacle sets are calculated for the moving object and its surroundings. To detect collisions the two obstacle sets are compared, searching for two or more equivalent cells in the path to be black. The representation by a matrix of spatial cells has the advantage that it is convenient for computer storage, but the computing time required to generate the representations of a moving robot tends to be large.

Khatib applied the method of representing obstacles by distance functions motivated by the electrostatic repulsion between like charges. For example, the mover and obstacles could be represented by positive charges. This artificial potential repulsion approach was aimed at the local, short-term avoidance of obstacles in real time rather than automatic planning of robot paths. Although the algorithm does not quite solve the find-path problem, the use of repulsion force made this algorithm original and the system worked in near real time. The function tended to infinity as the point approached the surface and was zero beyond a certain distance from the obstacle. This representation had the advantage that the task of calculating the distance between the robot and the obstacle was replaced by the task of evaluating the simpler function. Compared to solid geometry or polyhedral models, these calculations were relatively fast.

The repulsion force was generated by a fictitious potential field around each obstacle due to a potential assigned to it. When any link of the robot arm approached an obstacle, a repulsive force pushed the link away from the obstacle. If P was the potential function used, and D was a function of the minimum distance between the link and the obstacle, then P became large as D became smaller, and became zero beyond a preset distance from the obstacle. The force on the mover because of any obstacle was calculated from the equation

$$F = -\frac{dP}{dD}\frac{dD}{d\underline{x}}$$

where $\underline{x}$ is the position vector of the mover.

A higher-level planner was assumed to generate the initial path needed. Appropriate robot joint torques were calculated to follow the nominal trajectory and the force from the artificial potential field was incorporated to generate the final forces at the joints. This allowed each link of the robot to follow the nominal trajectory closely while avoiding any obstacles. The role of the artificial potential field was not to plan the path or the trajectory, but to bend it around obstacles. Khatib's algorithm was notable because the local obstacle avoidance problem was realised at the lower control levels for real-time execution, instead of being included in the path planner.

The main disadvantage was that only a limited number of obstacle shapes were available. Khatib stated "this potential is difficult to use for asymmetric obstacles where the separation between an obstacle's surface and equipotential surfaces can vary widely".

In the late sixties Pieper used a world model consisting of simple solid primitives (cylinders and spheres). Cylinders could be joined to form composite obstacles and spheres were assumed to be supported by planes. These models were approximate but simple. In the mid eighties this work was extended by Balding. He suggested that if the world were modelled just by spheres, intersection calculations could be greatly simplified. This is discussed in chapter four.

2.3 Modelling: The Robot.

Any robot consisting of a series of links and revolute joints may be represented by a general schematic model. If **n** is the number of robot joints, there are **n** coordinate frames which specify the robot configuration and variables which define joint positions in relation to the next.

This model has been widely used as the basis for modelling revolute robots and several engineers and scientists have described the process. The model defines the geometric relationship between joints.

At Stanford in the late seventies, Udupa simplified the geometric model of a modified Stanford Manipulator Arm to connected lines, then one line and then a point, by using obstacle transformations. Before any obstacle transformations were carried out the basic robot model was defined as two connected cylinders. The advantages of this representation were that path planning for a line or cylinder was much easier than the more complicated shape of the real robot. The method of modelling by connected cylinders was also used by Balding in the late eighties for a revolute robot.

Many other methods used polyhedral representations of the mover. This is a very accurate method of representing the robot but the computational effort for calculating collisions and path planning is so large that as yet it is impossible to use this depiction for real-time calculations.

An efficient geometric method using connected spheres was proposed by de Pennington. He was interested in collision avoidance rather than path planning. The method used a CSG solid model of the surroundings. The robot's path was simulated and the swept volume of the robot-sphere model calculated. The robot swept volume and the obstacle volumes were compared and where intersections between volumes occurred, collisions were indicated. The spheres produced swept volumes of regularized cylinders or "tori" under the restricted robot trajectories considered.

This method was unsuitable for automatic path planning and optimisation as the sweeping of the spheres was restricted to translational or rotational sweeping only. This excluded the movement of more than one joint at any one time and thus restricted the possible robot paths.

2.4 Path Planning.

The aim of robot path planning is to find a "trajectory locus" for a robot which will take it safely from one specified configuration to the next. The dichotomy of the problem is that the path produced should be as efficient as possible, but computer calculation time should be as short as possible.

Previous work in path planning has been governed by the obstacle and robot representations used. Usually, each path planning method may only be used with its own particular world model and robot model.

Most path planning research has not considered the case of open kinematic chains where all links in the chain must avoid collision. This situation is much more complicated than a free mover such as a mobile robot or unconstrained free moving shape. Often in the literature, the motion of the various simple movers considered has been restricted to pure translation, or to some mutually exclusive interleaving of translation and rotation. I will solve the automatic path planning problem for the case of an open kinematic chain.

Several authors divided path planning methods into two categories, local methods and global methods. Not all methods fit strictly into these categories but it is a useful categorisation and will be used in this section. A table showing the evolution of path planning techniques is shown in figure 2.1.

Local Path Planning Methods: Local methods use algorithms that find a path by repetitively considering configurations that are closer to the goal. When obstacles are encountered alternative strategies are tried, such as "Reverse and Move Left" or "Move Below". For local methods the problem is that of finding a series of intermediate positions connecting the **Start** and **Goal** configurations. The definition of the problem suggests that planning would be sensor based in any real time system. The advantage of these methods is that planning can take place when it is not possible to have a global world model. Local methods are often used in research for mobile robots and robots operating in unknown environments.

During his early work, Pieper and then later Balding utilised various heuristic procedures to move around a detected obstacle; for example, the arm folded to move in front of an obstacle and extended to move over an obstacle. If more than one obstacle existed, an ostensibly productive move which avoided one obstacle might cause a collision with another. This sometimes caused the manipulator to oscillate between obstacles. The avoidance routines sometimes generated non-productive moves so it was necessary to continually check that progress was being made towards the **Goal**. If no headway was being made, the path finding strategy was changed. In general the algorithms failed if paths led between obstacles, and in many cases there was no guarantee of finding a solution even if one or more existed.

Author	Date	Method	Obstacle Representation	Mover Considered
Pieper	1969	Local Planning	Cylinders & Spheres	Kinematic Chain
Udupa	1977	Local Planning	Polyhedral	JPLR (Modified Stanford arm)
Ahuja	1980	Local Planning	Polyhedral	3-D Octree
Lozano-Pérez	1981	Global Planning	Polyhedral	Polyhedra with restricted motion
Brooks	1983	Global Planning	Polyhedral	Polyhedra with restricted motion
Schwartz & Sharir	1983	Global Planning	Polyhedral	2-D Circles with restricted motion
De Pennington	1983	Collision Avoidance	CSG	Connected Spheres
Donald	1985	Global Planning	Polyhedral	Cartesian Robots
Canny	1986	Global Planning	Polyhedral	Polyhedra with restricted motion
Khatib	1986	Local Planning	Mathematical Functions	Kinematic Chain
Balding	1986	Local Planning	Spheres	Kinematic Chain
Tseng	1987	Global Planning	Polyhedra on floor surface	Kinematic Chain
Hwang	1988	Global Planning	Polyhedral	Polyhedra
Dupont	1988	Local Planning	Polyhedral	Kinematic Chain in a static cell

Figure **2.1**: Past Work in Robot Path Planning

Udupa meanwhile planned trajectories for the upper arm and forearm of the Stanford Manipulator Arm separately. Firstly a trajectory was hypothesised for the upper arm directly between **Start** and **Goal** configurations. Where collisions were detected, sub-goals were introduced which were intended to direct the path around the obstacles. For example, if a path between **Pos_A** and **Pos_B** was tested and a collision occurred then a sub-goal **Pos_C** between **Pos_A** and **Pos_B** was proposed. The paths between **Pos_A** and **Pos_C** and between **Pos_C** and **Pos_B** were then tested and so on, until either a clear path was found, or a calculation time limit was reached. Having found the upper arm path, the forearm was planned for positions where the forearm could collide with obstacles.

In the mid eighties Nguyen developed a fast heuristic algorithm for planning collision-free paths of a mobile robot in a cluttered planar work-space. The free space was described as a network of linked cones. Feasible positions and orientations of the mobile robot within the cone were computed. Feasible path segments were derived by local experts which used adjacency information of linked cones to generate local paths. Five local experts were used, namely, traversing a free convex region, sliding along an edge, circumventing a corner and going through a star-shaped region.

Khatib used the local method described in previous Sections. A manipulator moved in a field of forces. The obstacles were represented as repulsive surfaces and the goal as an attractive pole. The path planning method was to allow the summing of forces at each configuration to guide the robot to the goal.

This simple but effective method allowed obstacle avoidance to be carried out in near real-time using two PDP 11 computers and is the only method the author has found in the literature that worked in near real time. The method could become trapped in a local point of minimum force if the robot was drawn between two obstacles where either no possible path existed or the robot had to pass close to the obstacles. This restricted the method to very simple environments.

Lumelsky has recently developed an algorithm for planar 2-link manipulators which acquired the obstacle information from touch sensors located throughout their surface. The movement of the manipulator was restricted to those corresponding to linear changes in the joint variables, or those keeping parts of the manipulator in contact with the obstacles. When the manipulator hit an obstacle while travelling in the free space, it slid around the obstacle maintaining contact with the obstacle.

Even more recently Tseng used an Archimedes spiral to define a path for the lower joints of a T^3-776 robot. A path for the upper joints was then planned. All obstacles in the work-place were assumed to be resting on the floor of the work-cell and the upper arm passed over the top of the obstacles. The algorithms could not deal with obstacles which required the robot to pass under an obstacle.

Global Path Planning Methods. Global methods are usually applied after the problem has been reduced to finding a path for a point through space. This book will use the term "configuration space" for the 3-D quantised joint space. Since this is the natural space for the robot, different paths can be easily compared. The actual path planning takes place in the subset of configuration space through which the point may pass, Gouzenes called this "empty space" and it has also been termed "free space" by Brooks. I will use the term **free space**.

There are two global approaches for finding free space:

(a) Calculate the space occupied by obstacles and subtract this from the configuration space.

(b) Calculate the empty space directly.

The choice of approach depends on the type of representation used, and whether space is expected to be cluttered with obstacles or sparse. The fewer the obstacles, the more efficient is method (a) and the smaller the free space, the more efficient is method (b).

Lozano-Perez, Wesley and several other scientists assumed a fixed orientation of a body, and a planar case. Lozano-Perez, Wesley and Donald represented the surfaces of obstacles in configuration space. Except for very simple cases this is not easy to do. The simple cases used were 2-D and 3-D translation of solid objects. This amounts to a simple growing operation when the obstacles are transformed into a space they called C-space. When a rotation is allowed the transformation is no longer obvious. Mobile robots are the most practical example of this problem, usually possessing two translational degrees of freedom and a rotational degree of freedom. The problem is reduced to the motion-planning of a point amidst grown obstacles.

In subsequent work the space occupied by obstacles was calculated using a slice projection technique. Projections of the obstacles onto horizontal planes were calculated for a range of Z axis values in Cartesian space. These obstacles were then transformed into configuration space by considering the size and range of orientations of the moving object. Position and orientation of an object was represented by a six-dimensional vector or a point in configuration space.

The algorithm worked for cartesian manipulators only. Obstacles were polyhedral prisms whose axes were perpendicular to the horizontal plane.

Objects were modelled as trees of convex polyhedra and space was represented by a tree of full, mixed or empty cells. A graph containing the empty cells was defined by considering the connectivity of the cells. The cells made up a graph they called the visibility graph in which each node was the vertex of a polyhedral obstacle. The path planning problem was solved by the graph searching method developed by Hart in the late sixties.

Schwartz & Sharir extended the work of Lozano-Perez to path planning for several disjoint discs. They considered the task of moving several circles among polygonal obstacles in the plane, a collection of several line segments joined at a common point and finally a rigid rod moving in three dimensions. The work was not extended to a robot manipulator or kinematic chain.

Meanwhile Brooks implemented a planar path planner by modelling the free space between obstacles as generalised cones. Brooks modelled empty space as "freeways" along which the manipulator could move. He used a PUMA robot and separated the planning of the upper arm and forearm. The upper arm planning was done in joint space. Similar freeways were defined for the work-piece in real space. A path was found by considering the path for the upper arm and then seeing which work-piece freeways could be used with the upper arm path. Finally those paths which would cause a collision for the forearm were rejected.

This type of path planning produced paths which tended to have good clearances from obstacles. The method greatly restricted possible solutions because the constraints were applied to the movements whilst concurrently planning the upper arm and the work-piece in different spaces.

Chien used the concept of a rotation mapping graph (RMG) to plan paths for a rod, and then they extended the idea to cover the Stanford Manipulator Arm. Empty space was modelled as regions of collision free motion for the forearm. These regions were limited to those which implied collision free motion for the upper arm. These regions were then converted into a graph for searching by using a connectivity algorithm. Chien did not comment on the implementation of the algorithm or whether any practical work was completed.

Other scientists calculated the configuration space for the upper arm of the Stanford Manipulator Arm. The path planning for the upper arm consisted of planning a path for a point among a polyhedral representation of the configuration space. The shortest path for a point through this space was in straight lines between the edges of these obstacles. In later work, Luh presented an algorithm which, given an ordered set of edges, produced the minimum distance path. However, how to find which set of edges to use for the best path was not discussed.

O'Dunlaing studied planar motion problems using retraction to generalised Voroni diagrams and this work was further developed by Canny. A Voroni diagram of a set of obstacles represents the locus of points that are equidistant from at least two of the obstacle surfaces, that is the locus of maximally distant points. Searches of the Voroni diagram tend to give the safest path solution in terms of distance from obstacles. An example is shown in figure **2.2**.

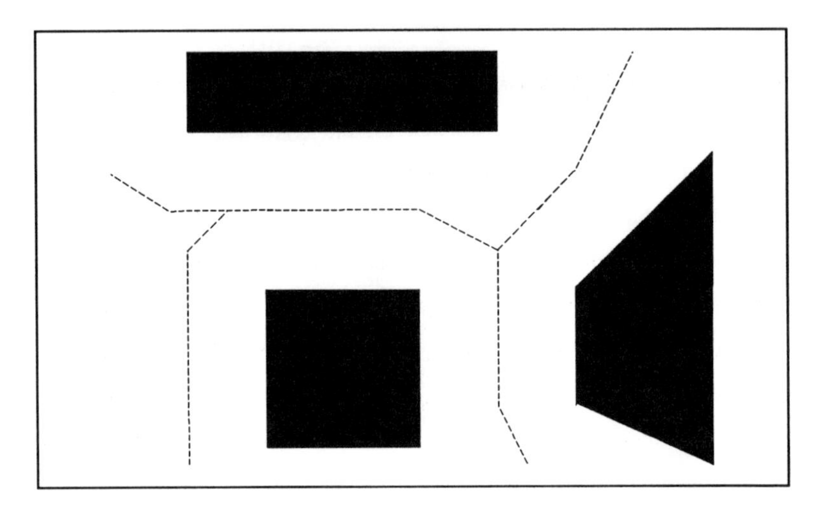

Figure 2.2: An example of a Voroni Diagram.

The problem of moving a ladder or a line segment among rectangular obstacles was considered by Maddila. The global problem of moving a ladder was decomposed into several local motion planning problems. The free space was divided into corridors and junctions. Corridors are the hallways between rectangular obstacles, and junctions are the areas where corridors meet. The movement of the ladder was either horizontal or vertical, and rotations took place at L-shaped junctions. A weighted graph called a motion graph was constructed from the solutions of the local sub-problems. The weights represented the longest ladder that could be moved between the nodes of the motion graph so the algorithm was also capable of finding the longest length of the ladder that could be moved between two positions in the free space.

Lozano-Perez and Wesley had used a similar map of free space called the visibility graph or V-Graph during the late seventies. A V-Graph connects those vertices of obstacles that can be connected by a straight line that does not penetrate any obstacle. This is referred to as being able to "see" another obstacle. This is shown in figure 2.3.

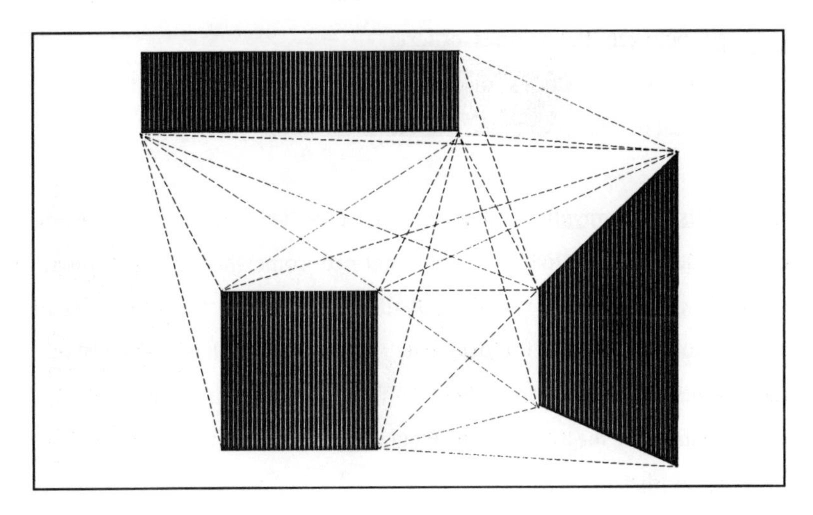

Figure 2.3: An example of a Visibility Graph.

2.5 Path Optimisation.

A few path planning algorithms have considered robot kinematics but hardly any have also considered dynamics. Research work considering system dynamics has not yet achieved on line automatic programming of robots.

Path optimisation algorithms used in the past have attempted to minimise some cost function of a robot path. Different criteria have been considered when estimating the cost. Five of the most important criteria are

(i) Distance travelled.
(ii) Time taken.
(iii) Energy used.
(iv) Component wear.
(v) Path safety.

The weighting given to each criterion in assessing the cost of a path must be decided before path optimisation can take place. Different applications require different emphases to be placed on the different criteria. Some criteria may oppose each other, for example, as the time taken for a robot path decreases so the energy consumed tends to increase. The "optimum path" is the compromise required between the various criteria.

The following Sections discuss the factors affecting the five criteria and their effect on each other.

(i) Distance travelled. The distance travelled by a robot may be defined, either as the distance travelled by a point defined somewhere on the robot (usually the gripper), or by the total amount of movement which the robot has made.

The total amount of movement which a robot has made is the sum of the movements of each robot axis. If there is a mixture of linear and rotary movements then the rotary movements may be converted to linear ones by defining their linear distances as,

$$L_d = \theta \times L$$

where θ =the rotary movement
L =the length of the link
L_d =the linear distance

Several scientists and engineers used the total distance moved by a robot to calculate their "minimum distance" paths. Gilbert & Johnson used the distance travelled as a cost. Their solution technique employed an interior penalty function, later used by Dubowsky. When a robot is programmed to move between two positions there are many different ways it may move, but most robots move in one or more of three different ways:

(a) Independent movement of axes
(b) Point to point linear interpolation
(c) Interpolation of robot axes.

It is important that any path optimised for shortest distance does not oppose the method used. The methods are described:

(a) Independent movement of axes. The robot's axes move independently from their starting positions to their finishing positions. This type of movement requires the minimum of computer control but it is difficult to model.

(b) Point to point linear interpolation. A point is defined on the robot, and the robot moves such that the point travels in a straight line from the start point (START) to the goal point (GOAL).

(c) Interpolation of robot axes. The robot's axes are interpolated such that they all have the same function of time. For example, if one of the robot's axes has initial value θ_a and final value θ_b then,

$$\theta(t) = \theta_a + f(t)(\theta_b - \theta_a)$$

and f(t) is the same for all other axes. For this type of interpolation all points on the robot arm describe complex curves in three-dimensional space.

(ii) Time taken. The time criterion has tended to be applied to trajectories produced by path planners and not by the planners themselves.

In a study of minimum-time manipulator trajectory planning, Luh & Lin constrained Cartesian velocities and accelerations. Their scheme required experimental identification of Cartesian velocities and acceleration bounds. Kim & Shin, in a similar study, developed a method for minimum-time trajectory planning in joint space. In their study, an absolute path deviation was prescribed at each corner point, and local upper bounds on the joint accelerations derived from the arm dynamics.

Several computer scientists applied dynamic programming to the planning of trajectories where the path was specified, the control forces/torques bounded and the travel time given. Bobrow and others devised a specific technique to solve minimum time trajectory planning problems for a manipulator following a prescribed path under state dependent constraints on the torques/forces. Their algorithm cannot be extended to other performance criteria.

The time for a robot to move from one position to another depends on the following.

(a) Path Length.
(b) Path Complexity.
(c) Path Type.

(a) **The path length**. The greater the path length, the longer the minimum time taken for that path. The minimum possible time for a path is assumed when at least one robot joint is always changing at a maximum rate during the path. The speed of the robot's path is in turn affected by the complexity of the path.

(b) **The Path Complexity**. For a complex path, a larger amount of time is spent in accelerating and decelerating the robot arm so that average velocity is reduced.

(c) **The Path Type.** Paths which require large amounts of computing time to calculate, such as point to point linear interpolation, take a longer time to execute than paths calculated by, for instance, the interpolation of axes.

Sahar & Hollerbach recently described a general method for the planning of minimum time trajectories for robot arms. Sahar reports that 'optimal paths tend to be nearly straight lines in joint space'.

(iii) Energy used. During the seventies, Paul presented a technique which allowed the manipulator to transit smoothly from one straight-line segment to another. The motion was continuous in joint displacements, velocities and accelerations. Others used the criterion of energy used by the robot motors to create a cost function for the robot path. They found that the factors affecting the energy used were the following:

(a) Distance Travelled,
(b) Time Taken and
(c) Path Shape.

(a) **The distance travelled.** Energy is dissipated in friction as the robot moves, so the further the robot moves the more energy is required.

(b) **The time taken.** As transit time decreases for a given path, so accelerations and decelerations for the robot increase. This increases the energy used.

(c) **The Path Shape.** Smooth paths require least energy because accelerations and decelerations are reduced. Figure **2.4** shows the minimum distance path for a point moving around a rectangular obstacle. Figure **2.5** shows a path which would use less energy.

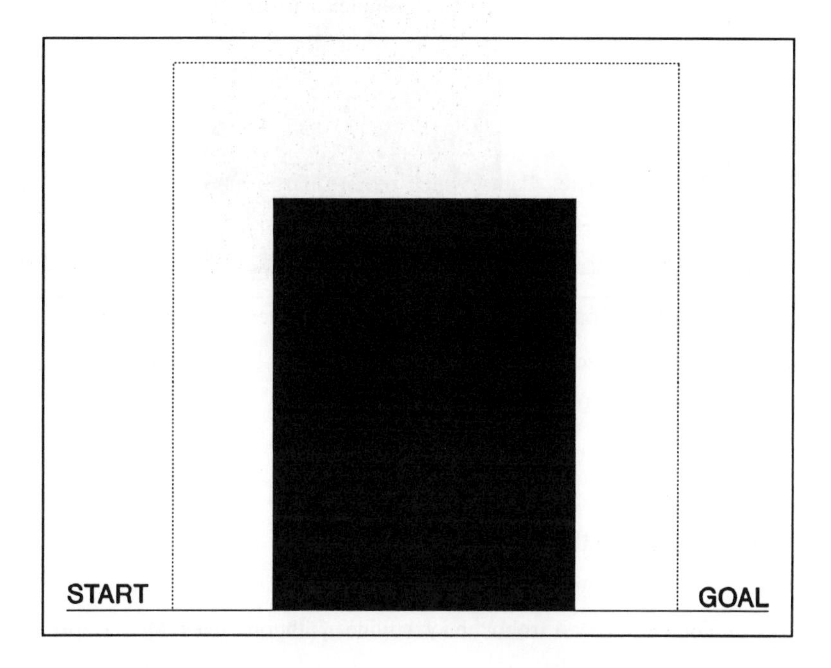

Figure **2.4**: Minimum distance path

(iv) **Wear on the components.** During the eighties I was busy considering wear and forces on the structure and the components. As wear affects the mean time between failure and the servicing interval for a robot, reducing wear will increase productivity and reduce operating costs .

Wear on robots is affected by the same factors that affect the energy used; that is, distance travelled, time taken and the path shape.

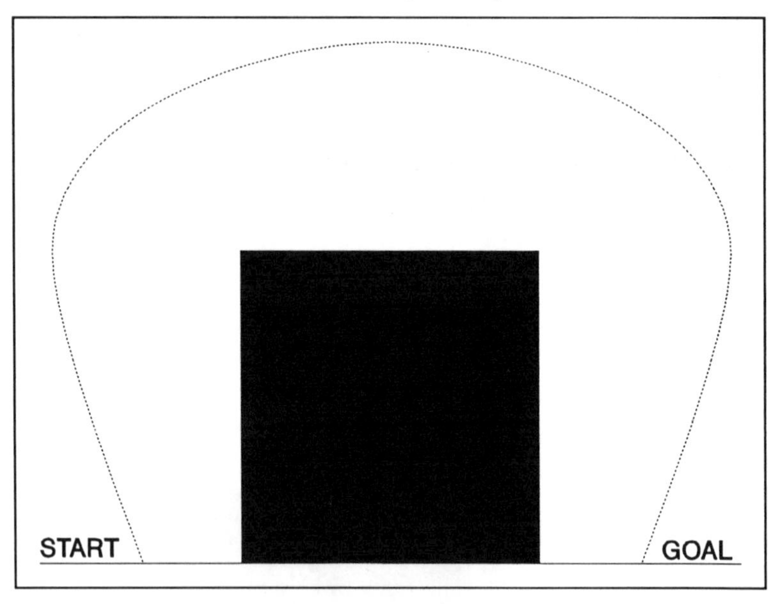

Figure **2.5:** Path requiring less energy

(v) Safety. Bonney described how the safety of a path may be viewed from three different standpoints.

(a) **The robot.** A robot may collide with obstacles if it is programmed to move too close to them. The path along which a robot is programmed to move may be different to that which it actually takes. One particular problem is the rounding off of corners. Most robots will follow straight paths which blend into other straight paths unless they are programmed to wait at via points. To reduce the danger of a robot hitting obstacles, the nominal size of the obstacles may be increased by some safety margin. This ensures that if a robot does cut corners it will still miss obstacles.

(b) **The work-piece.** If the robot is moving quickly the forces on the work-piece will increase. This may cause the work-piece to move in the gripper or be dislodged from it.

(c) **Humans.** As the speed of the robot increases so the danger to human operators is increased. This means that additional safety precautions may have to be taken.

Chapter Three

DEVELOPMENT OF A TYPICAL SYSTEM

3.1 Introduction.

This chapter describes the development of typical equipment and representative software systems which could run on the apparatus.

Figure 3.1: An example of motion planning apparatus.

Most modern and complex industrial manipulators use sampled data control systems with hierarchical structures so the development of such a system is described. An example of a final apparatus is shown in figure **3.1**. It consists of a camera which provides an input to a computer vision system. This is connected to a path planning computer. A third computer is a dedicated robot controller with associated interfacing and DC servo-amplifiers, connected to a robot.

On such a test rig the following processes need to be implemented:-

(a) Motion Planning.
(b) Robot Control.
(c) Vision data acquisition.
(d) Path Improvement.
(e) Image data processing.

A simplified block diagram of a final system is shown below in figure **3.2**.

Development of suitable sub-systems within the main computer and the controller are described in the following sections. Vision systems are described in chapter five.

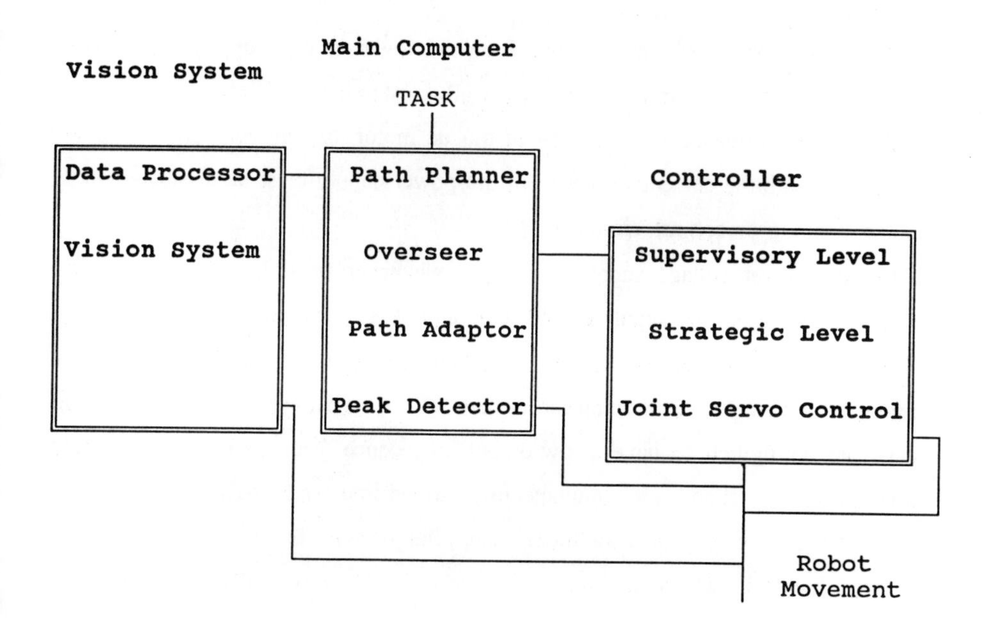

Figure 3.2: A Simplified Block Diagram of a Final System

3.2 The Development of an Apparatus: A Servo Amplifier

A suitable actuator is required for the machinery. DC-motors are common drivers for complex programmable machines, including robots, and these require DC-servo amplifiers. A novel amplifier design is described in this section which is powered from more than one voltage supply and solves some of the common problems associated with servo amplifier design. It is common for amplifiers to provide current to the d.c. motor in a forward or reverse direction from a single power supply. In the amplifier described here current is applied by four separate high impedance output stages. The design is highly efficient and largely overcomes the problems of crossover distortion and wasted power common to conventional servo motor power amplifiers.

For rapid speed and fast responses, the amplifier must be capable of delivering a substantial current. In this novel design, large demand voltage signals have current supplied from a higher voltage positive or negative supply and this is shown in figure 3.3. This would occur at high speed or for torques associated with large changes of force.

For smaller inputs, current is drawn from two lower voltage supplies. This would be the case when the motor is at rest while sustaining a constant load. When stationary the mechanical efficiency of the dc motor is near zero and the power associated with the current drawn from the power supply must be dissipated in the motor winding or control circuitry.

Because lower voltage supplies are used whenever possible, a power saving is achieved. Considering supplies ±40 volts and ±10 volts, the thermal dissipation of power in the controlling circuits is less than one eighth of that in a conventional circuit of ±40 volts. The amplifier circuits of a conventional twin supply output stage are configured to have a low output impedance and are liable to excessive common current if both are simultaneously driven into conduction. This transition characteristic was crucial in traditional amplifier design. Basic design provides for a 'dead band' in which no conduction occurs. Crossover distortion was then present in the output waveforms.

A mixer stage is required to provide an error input for the novel amplifier. This can take the form of a non-linear circuit based on a TLO81 operational amplifier. A tested circuit is shown in figures 3.4 and 3.5. The TLO81 has proved itself as a robust and reliable op amp for control applications, directly replacing standard op amps such as the 741. The mixer has two feedback paths, one of which is non-linear and only takes effect within predefined limits.

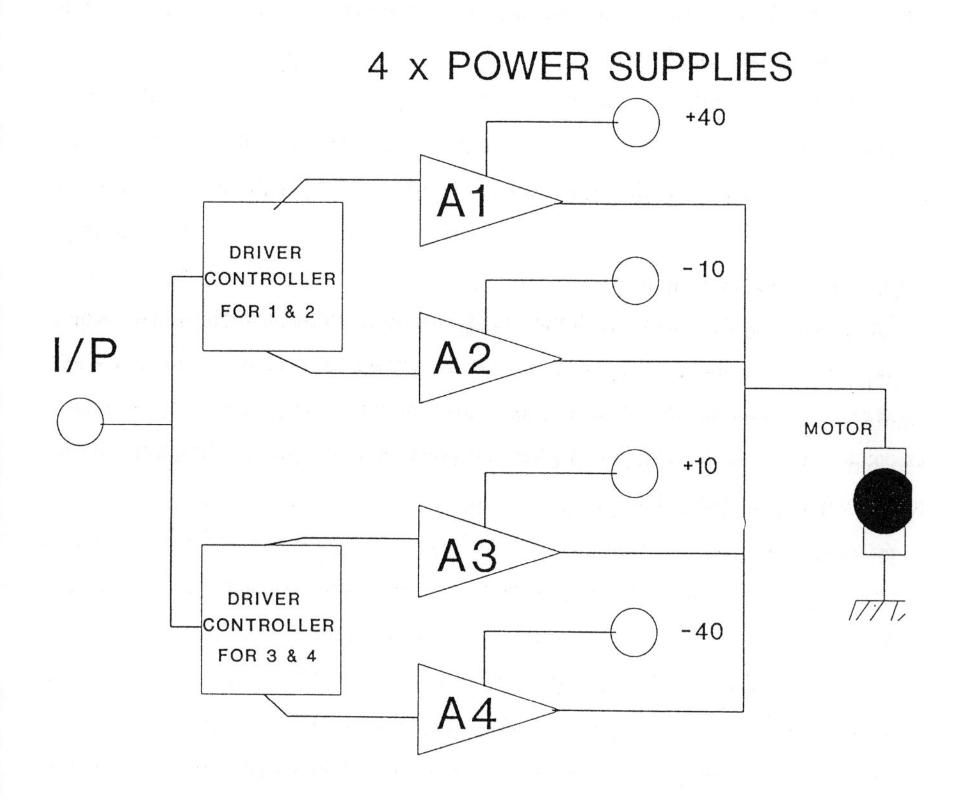

Figure 3.3: Block diagram of a novel servo motor amplifier

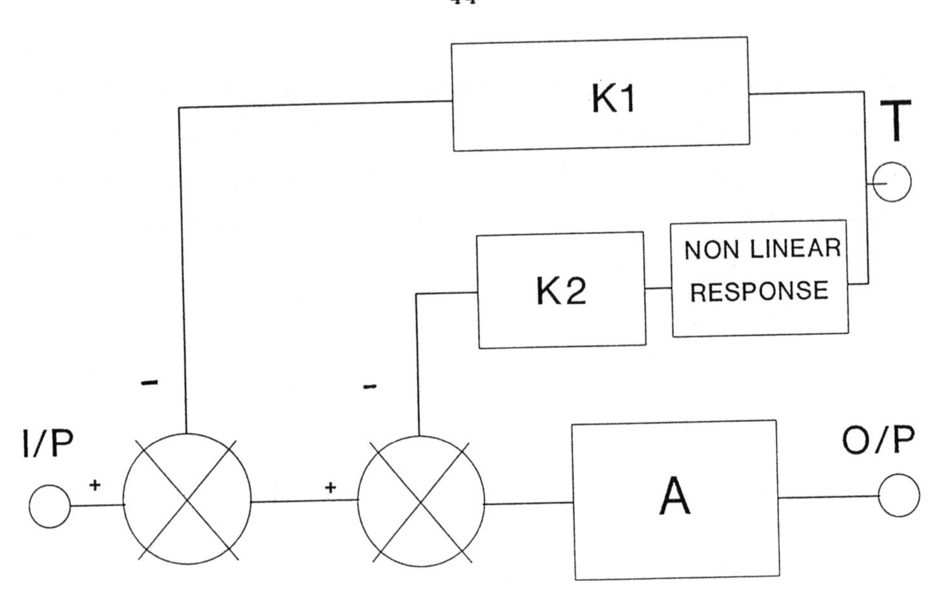

Figure 3.4: Block diagram of a mixer circuit

The block diagram of the mixer circuit is shown in figure 3.4. The feedback signal is split into two paths, both providing negative feedback. The outer loop is a simple linear feedback loop, but the inner loop is non-linear. The gain and range within which the non-linear circuit takes effect can be preset by the selection of suitable resistors and voltages. The outer loop is "loose" (low gain), allowing high speed. The inner loop is high gain, providing "tight" control within the limits.

The demand input to the mixer is an analogue voltage from the controller computer derived from a D/A circuit. The tacho signal is optional and can be used when a speed/voltage signal is available. The tacho input is mixed with the demand input and fed to a TL081 connected as a standard mixer using the negative input.

A circuit diagram of such a mixer circuit is shown in figure 3.5. In this circuit the tacho signal is separated into two paths. One path is high gain, for low velocities close to the demanded position and consists of the four diodes MD1 - MD4 and the three resistors MR3 - MR5. This inner loop only takes effect when the tacho-feedback signal is within the two supplies at A and B.

A current will flow from supply A to supply B. When the tacho input T is zero, half the current will be flowing through MD2 and MD4, and half through MD1 and MD3. As the tacho input moves away from zero, say positive, the amount of current through MD1 and MR3 reduces and current through MD3 and MR5 increases. Thus, the voltage applied to the op-amp tends to increase in sympathy. The effect within the range is that the output to IC1 is via MR4.

The high gain circuit only takes effect within the limits of A and B, since when the tacho input is outside this range, say positive, no current can flow through MD1, which will be reverse biased. So in this circuit MR2 and MR4 define the inner loop gain, and in this case a loop gain of 42/5.6 =7.5.

A second path with a low gain for higher velocities where control is not so important, consisted of a simple resistor, MR6. A capacitor, MC1, is also included to remove noise from the feedback signal. MR6 and MR2 define the outer loop gain, and in this case the loop gain is 43/56 =0.77.

The control circuits of the amplifier can be configured to have a high output impedance, giving an output current which during the conducting phase of each circuit varies linearly with the applied demand signal. These outputs can be safely connected. The lower voltage circuits can be biased so that over a central range of input control signal, both circuits conduct. Outside the range, one or other circuit is cut off as shown in figure 3.7. Within the range, the rate of change of output current with respect to the input signal will be twice that outside the range and neither dead band nor discontinuity of current will occur.

The four circuits can be configured so that the higher voltage amplifiers only conduct once the opposing low voltage amplifier has turned off. A typical amplifier circuit diagram is shown in figure 3.6.

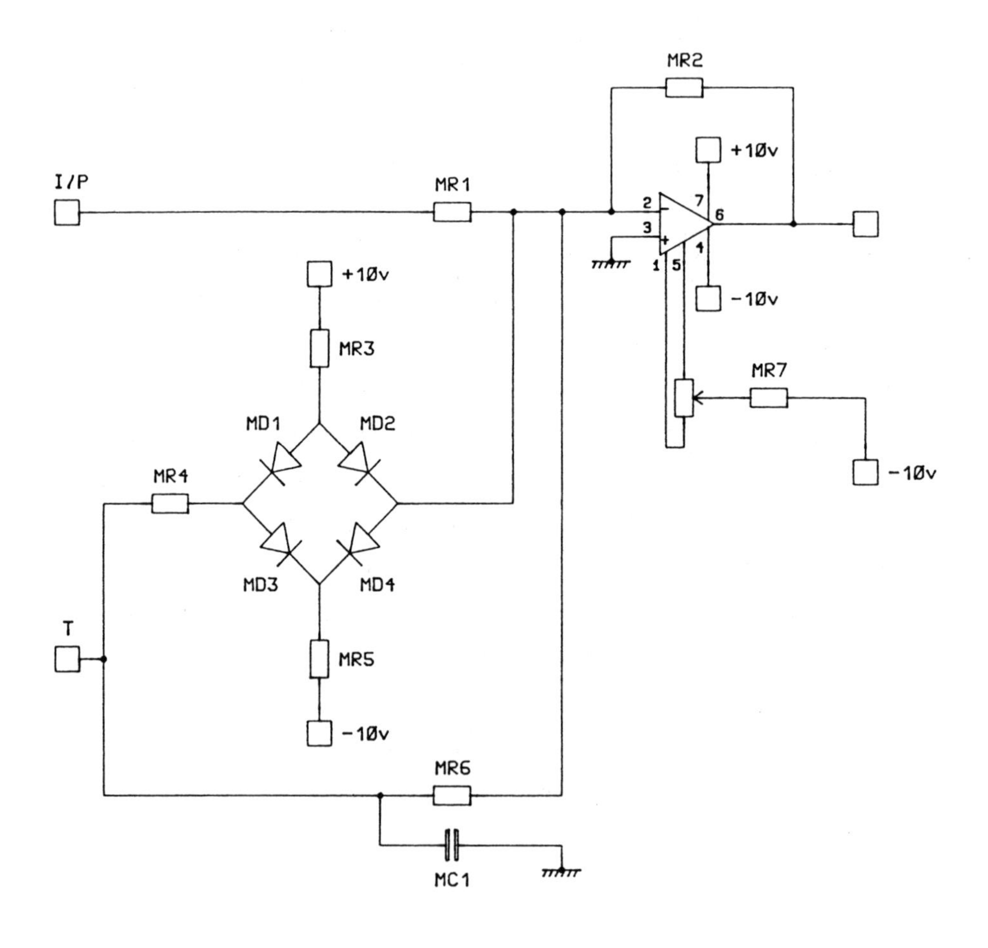

Figure 3.5: A tested circuit diagram for the mixer stage

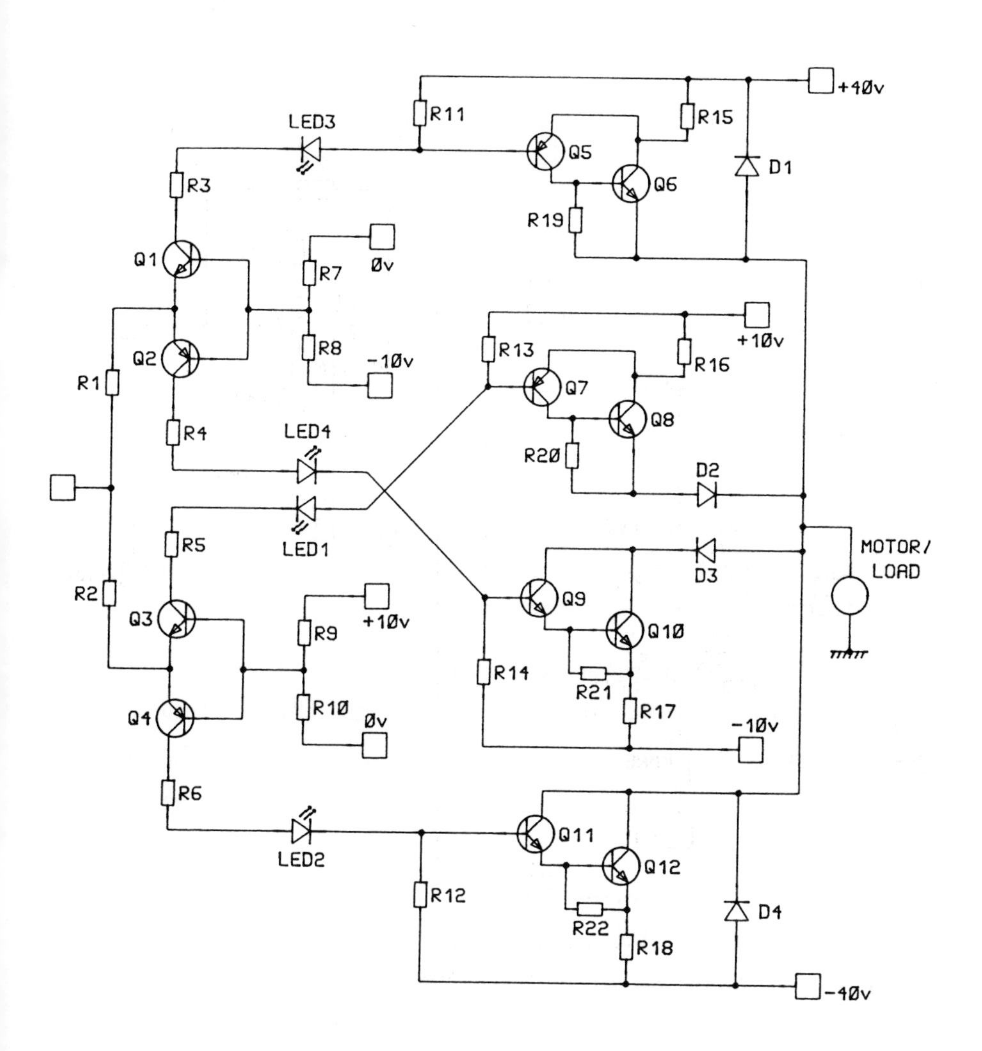

Figure 3.6: Circuit Diagram of the Servo Amplifier

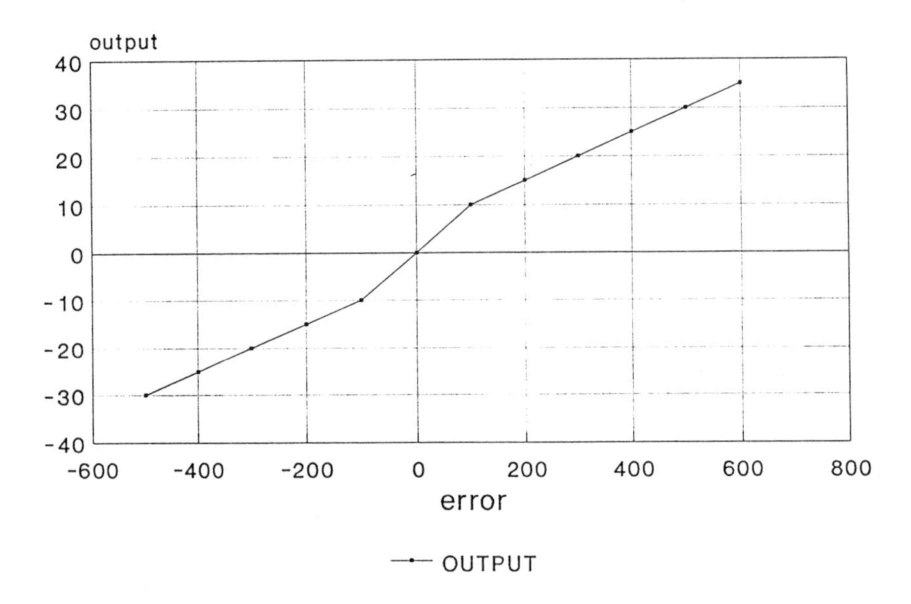

Figure 3.7: Error/Output

3.3 The Development of an Apparatus: Computers and a Robot

Intel 8086 and Intel 80286 based microcomputers are relatively cheap and easily available. These computers are the minimum standard needed to allow the path planning and path adaption algorithms described later to be implemented. Communication between these computers is discussed later in section 3.5.

Many suitable robots are available including Unimation Puma, Syke Robotics 600-5 and Mitsubishi RM.501. The Unimation and Syke robots are complete systems, but the Mitsubishi can be purchased as a stand alone piece of equipment. The mechanical structure, without a controller or servo-amplifiers, can be purchased. This allows access to the lowest levels of circuitry and machinery so the Mitsubishi RM.501 was selected for research work completed by the author at Portsmouth Polytechnic. The robot motors in a Mitsubishi are relatively small and they require correspondingly smaller servo amplifiers. The servo amplifier and mixer circuits described in the last section are suitable but must be redesigned for use with +/-24volts d.c. and to supply a smaller current. The Mitsubishi robot does not have integral tacho-generators, so a velocity signal for this type of machine must be derived in the hardware from the back EMF across the motor.

The controller can be equipped with a bus system such as the G64 bus to allow for system expansion. A typical set of apparatus excluding a vision computer is as shown in figure 3.9.

3.4 The Software Systems.

A basic system has three main levels:

The Supervisory Level.

The Strategic Level.

The Joint Servo Controller.

This is shown in figure 3.8. The software for the higher levels is generally written in a high level language and assembly language may be used for the lowest level.

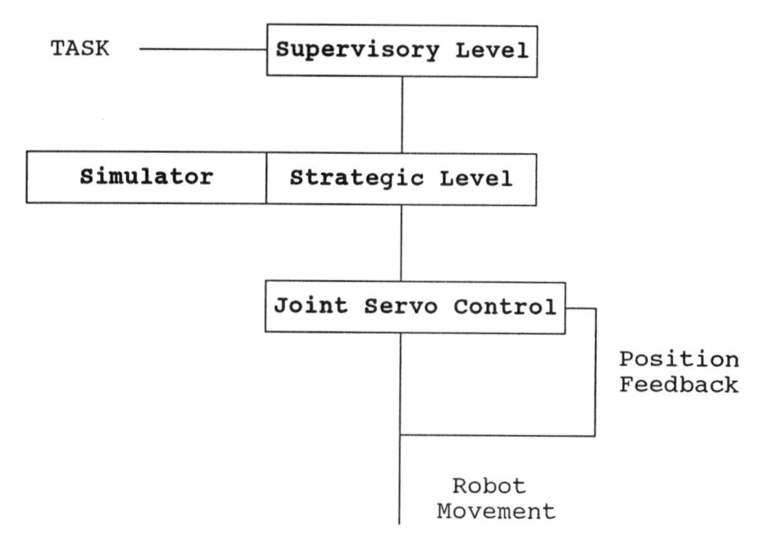

Figure 3.8: A simple control system for a single actuator.

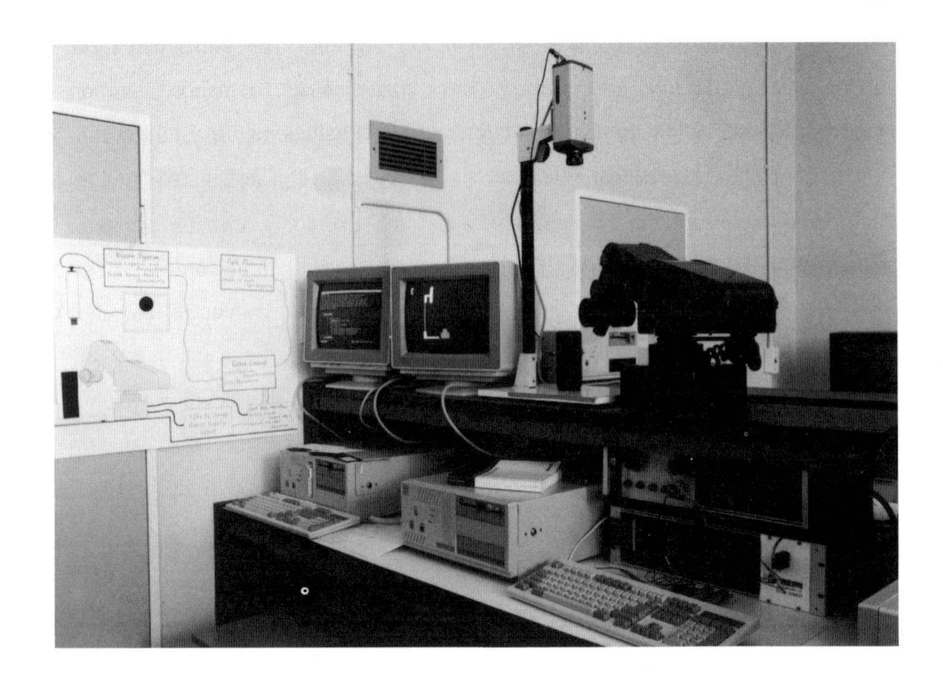

1. Camera
2. RM-501 Robot
3. Obstacle
4. Path Planning Computer
5. Robot Control Computer
6. Power Supply Unit
7. G64 Rack
8. PC LabCard Connector Card

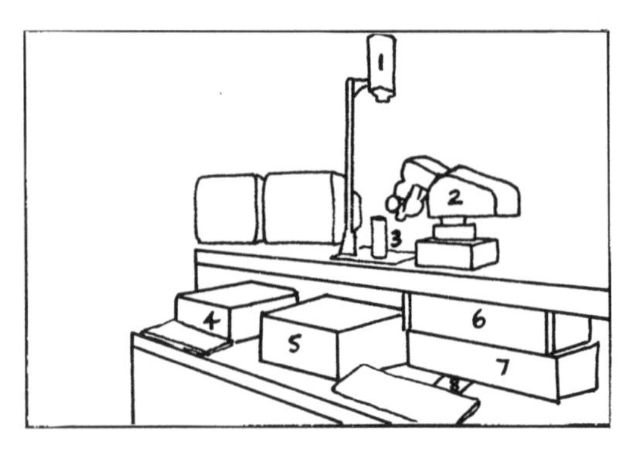

Figure 3.9: A set of apparatus (Excluding a vision computer)

(a) **Supervisory Level.** The supervisory level is the overall controller. This level handles interfacing with the human operator and the files containing the required movements of the robot joints θ_1 etc, and for any simulated joints $\theta_1 sim$ etc.

(b) **Strategic Level.** The strategic level considers the demanded motion and assigns look up tables containing output voltage values to be used by the joint controller level. Demanded elementary movements are distributed to the joint control level.

(c) **Joint Servo Controller.** This level is a dedicated joint controller for each joint, θ_1, θ_2 etc, and realizes the functional movements by controlling the joint angles using a position servo.

A Peak Detector can be added to this system. This is a low level program which samples the currents to the joint motors. In a simple system the Peak Detector can feed information directly to the Supervisory level as shown in figure **3.10**.

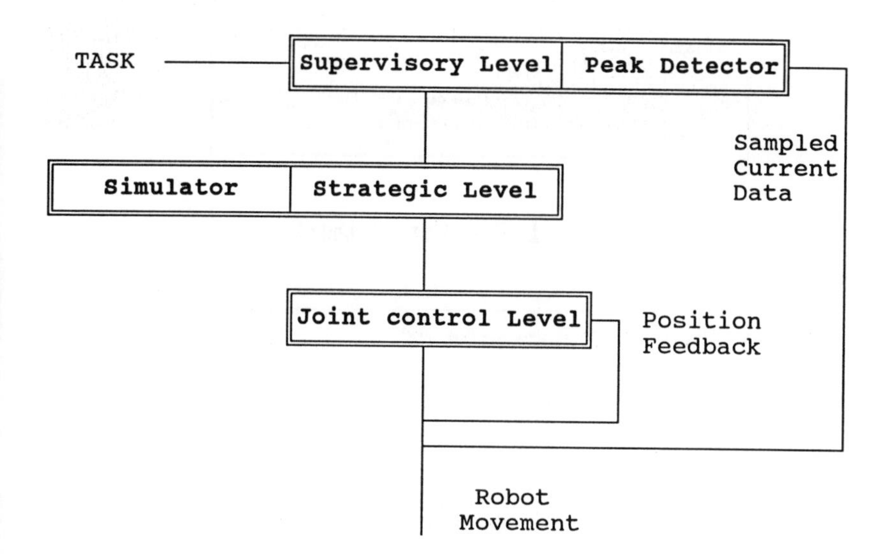

Figure 3.10: The addition of a Peak Detector.

With the robot in motion, the current to the base motor from the DC-Servo Amplifier can be recorded for a variety of velocities, loads and accelerations. The motor current can be sampled via an A/D converter and will be typically as shown in figure **3.11**. This waveform is too noisy for interpretation, so smoothing methods must be investigated. Suitable results can generally be obtained using a simple low pass filter and a sample waveform is shown in figure **3.12**.

A study of the filtered wave-forms can suggest a new strategy for path adaption and force detection. This will be discussed later in the book.

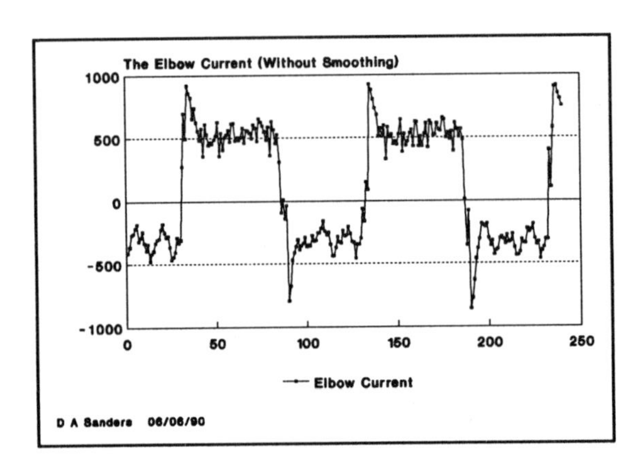

Figure 3.11: Raw Current Data

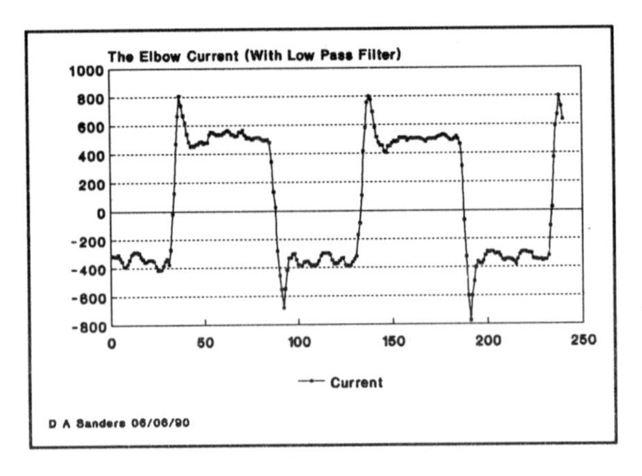

Figure 3.12: Current Data passed through a Simple Low Pass Filter

A novel parallel and hierarchical system structure is described which evolved from these systems. This is shown below in figure 3.13.

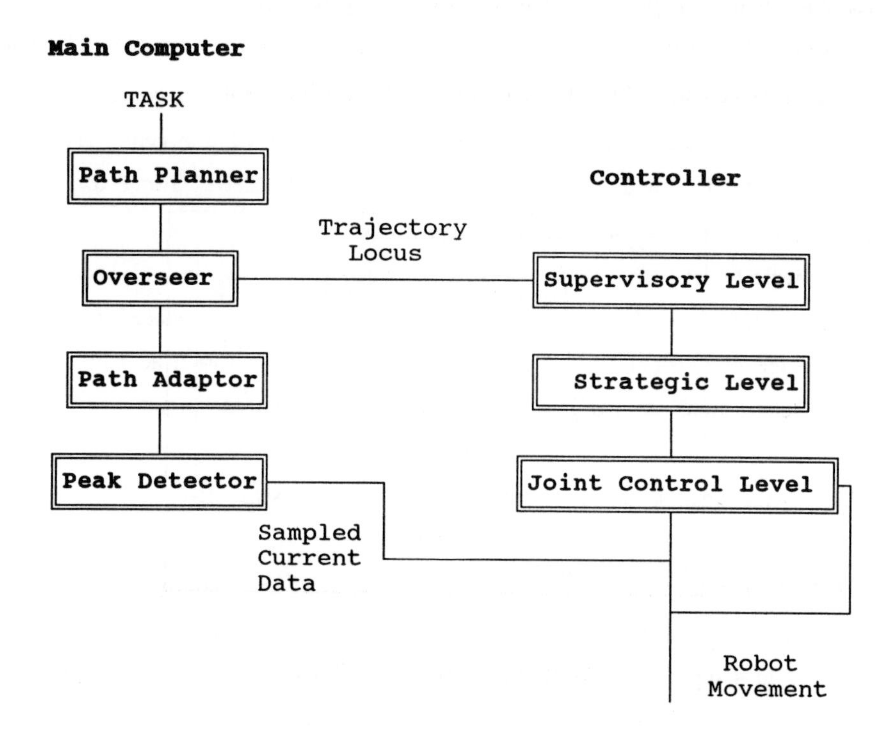

Figure 3.13: A Parallel Hierarchical Control System.

To improve the speed of data processing and avoid redundancy, decision processes can be performed in two computers and this is shown above. Slower control operations are established at the top of the structure and progressively faster operations can be performed at lower levels. High level decisions and improvement strategies are considered by the main computer while the second computer is controlling the robot and machinery. In both computers, after obtaining information from a lower level, each level can make decisions, {considering decisions from higher levels}, and forward commands to lower levels.

All levels and both computers are ruled by the path planner in the main computer. Target points cannot be passed to the robot controller until a path has been planned. The robot controller has layers or levels which are similar to the hierarchical system depicted. These are described below:

- **Supervisory Level.** In this new structure this level is no longer the overall controller as the path planner in the main computer rules the system. The level no longer interfaces with a human operator, but instead receives a trajectory locus from an overseer in the main computer.
- **Strategic Level.** This level works as described previously.
- **Joint Control Level.** As before, this level executes the imposed motion of each degree of freedom. Similar software is associated with each actuator except that different gains are required in the software loops for each joint.

The hierarchical structure of the main computer consists of:-

- **Path Planner.** The path planner accepts a START configuration and a GOAL configuration from a human operator and produces a trajectory locus which is passed to a supervisory level via an overseer. Typical algorithms are described later.
- **Overseer.** The overseer works in parallel with the controller and oversees the well-being of the robot. Specifically the overseer accepts a path description as a trajectory locus from the path planner or new path information from the path adaption level. This information is passed to the controller. During the adaption work described later in the book the motor drive currents are monitored for collisions at this level.
- **Path Adaption.** This level considers path adaption possibilities from the information provided by the peak detector - specifically, improvements to reduce the current peaks in the motor drive current.

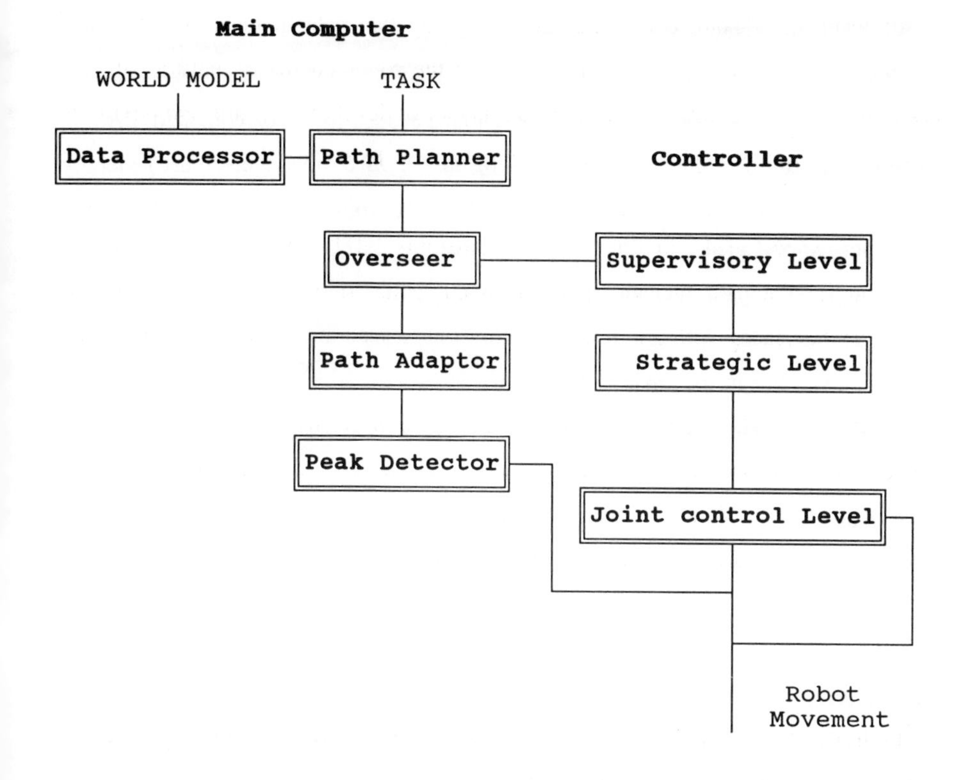

Figure 3.14: A typical complex system (Excluding the vision sub-system)

- **Peak Detector**. This is a low level program which samples the currents to the joint motors and considers the relative amplitudes of successive readings. If the current rises above preset limits, an interrupt routine informs the Overseer.

Given a task in the form of a start and goal position, the system automatically plans a path. In the later work described in this book, obstacles are recognised using a vision system. In a less complex system obstacles can be introduced into the main computer using simulated data stored on a disk. This requires the addition of a Data Processing level in the main computer.

The Data Processor processes the obstacle data and passes a list of blocked joint positions to the Path Planner. This level is one of the two main software modules "TransformSphere.BAS" or "TransformSlice.BAS" which are described later in the book in chapter five.

- **Software**. The main programs, communication routines and man/machine interfacing can be written in a high level language such as 'C' or compiled BASIC. Time critical routines and routines accessing the BIOS at low levels, must generally be written in assembly language. The assembly routines can be linked by a common variable syntax.

The high level programs are described later.

3.5 Communications.

Communicating between two computers is relatively simple and can be achieved using a single RS 232 serial link or parallel CIO 'plug in' boards.

For a more complicated structure including a vision system, a more intricate communications system is required. Each of three computers must communicate with each of the other two and to interrupt at various levels. The vision system must interrupt the Path Planner and Controller if an obstacle appears within the work place. The robot must interrupt the vision system to inform it if the robot is about to pass under the camera.

A simple method is to use spare ports on plug in LabCards. An alternative is to use two serial RS-232 ports in each computer to permit the use of port interrupts with "On Event" instructions. IBM compatible micro-computers are generally fitted with a 25 pin port as COM1 and/or a 9 pin port as COM2.

The main sets of data to be transferred are joint angles describing configurations of the robot. Blocked configurations are passed from the vision system to the Path Planner and the trajectory locus is passed from the Path Planner to the Controller. Other data which may be handled by the communications sub-system includes:

(a) A signal from the Vision System when a change in the environment is detected. This code may also be passed to the Controller to warn that a new trajectory locus is being prepared. After receiving this code the Controller must avoid moving the robot into the work area.

(b) An end of file code. If there were no blocked nodes then the Start of File code can be immediately followed by this code.

(c) A warning to the Vision System when the robot is about to enter the work cell. This will stop the Vision sub-System from detecting the robot as an obstacle. When the robot is moving away from the work cell an "All Clear" code can be passed to the Controller.

The "All Clear" signal depends on the number of dark pixels detected by the vision system. To test if the work-cell is empty, the number of black pixels in a frame can be tested against a preset background noise level. If the frame is empty the all clear flag can be set to TRUE and the Start of file, End of file codes passed to the Path Planning computer.

3.6 The Inclusion of a Bus System.

The inclusion of a bus system will allow for the expansion of the hardware and software systems in the future. A range of industrial modules are available for various bus systems. The G64 bus is an industry standard and this will be described. The bus system consists of the following:-

(a) A Transmitter card which connects to the robot control computer.

(b) A G64 rack system with an independent power supply.

(c) A ribbon cable and Receiver card which converts the 80286 machine signals to G64 specifications and connects to the G64 rack.

The G64 bus is a 16 bit bus with 17 address lines capable of addressing 256K. The system is shown in figure **3.15** with the servo-amplifiers mounted above. Computer interfaces are available or can be designed for the dc servo-amplifiers to plug into the G64 rack. Interface cards can typically contain DACs to convert the digital output from the computer to analogue voltages for the servo amplifiers and decoder circuits for the optical encoders.

1. ±10VPower Line
2. DAC Input Connector
3. Optical Encoder Power Supply and Connector
4. DC Servo Cards
5. Motor Power O/P Cable
6. G64 Ribbon Cable
7. G64 Receiver Card
8. G64 Cards Containing DAC s and Optical Encoder Decoders
9. PC LabCard Input/Output Connector Card

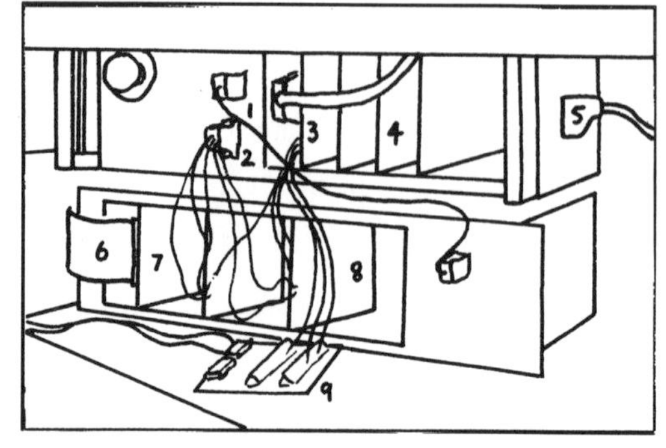

Figure 3.15: The G-64 Bus System and Servo-Amplifier rack.

The table below shows some tested locations in memory for G64 interface cards.

G64 Memory Addresses		Optical Encoder Input Location	
Joint	**DAC Location**	**Hi-byte**	**Lo-byte**
Base	&AF404	&AF400	&AF401
Shoulder	&AF406	&AF402	&AF403
Elbow	&AF504	&AF500	&AF501

If LS2000 chips are used then these give a two byte representation of each joint angle.

3.7 An Example robot.

A Mitsubishi RM.501 robot will be described as an example. The robot is a five degree of freedom manipulator with three links as shown in the GRASP plot (figure **3.16**). The main disadvantage in selecting this robot was the small working envelope and this is discussed further in chapter five.

The first three joints will be considered in this book and the solution to the position problem for this robot is described. The robot link lengths are:- Upper-Arm =L1 =220mm. ForeArm =L2 =160mm.

Three points on the robot were considered:- The Origin, the Elbow and the Fore-Tip.

(a) The origin of cartesian coordinates was set at the centre of the shoulder joint θ_2 and the base joint θ_1. This point was referred to as "Origin".

$$Origin_x = 0$$
$$Origin_y = 0$$
$$Origin_z = 0$$

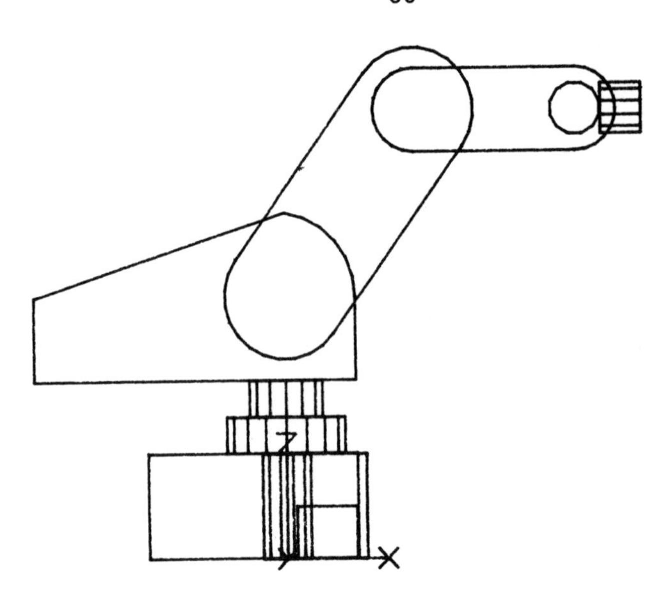


```
ROBOT  RM501 NEW TYPE XYZ_7004          JOINT  1 (SHIFT  Z 166 ) REVOLUTE  Z
                                                  JOINT  2 (SHIFT  Z  80 ) REVOLUTE  Y
                                                  JOINT  3 (SHIFT  X 225 ) REVOLUTE  Y
                                                  JOINT  4 (SHIFT  X 160 ) REVOLUTE  Y
                                                  JOINT  5 REVOLUTE  X
                                                  TAP (SHIFT  X 65 )
                                                  MINIMUM  -150 -65 -45 -90 -180
                                                  MAXIMUM   150  65  45  90  180
                                                  VELOCITY  400 400 400 400 400
                                                  ACCELERATION   160 160 160 160 160 ;
CUBOID  BASE 208 180 44;
TO RM501  ADD BASE (SHIFT  X -137 Y -90) ;
CYLINDER  %PED1 LENGTH  100 DIAMETER  120 TOLERANCE  1 ;
CYLINDER  %PED2 LENGTH  22 DIAMETER  75 TOLERANCE  1 ;
SET PED =        %PED1  %PED2  (SHIFT  Z 100)
TO RM501  ADD PED (SHIFT  Z 44)
POLYPRISM  %PED3 HEIGHT  120 AXIS Y      0      0      320      0      320      80
ARC  TO 250 160 RADIUS  -80 TOLERANCE  1  0      90 ;
TO RM501 _J1  ADD %PED3  (SHIFT  X -250 Y -60 );
POLYPRISM  ARM1 HEIGHT  100 AXIS Y      0      0      225      7.5
ARC  TO 225 92.5 RADIUS  -42.5 TOLERANCE  1             0      92.5
ARC  TO 0 0  RADIUS  -50 TOLERANCE  1 ;
TO RM501_J2  ADD ARM1  (SHIFT  Y -50 Z -50 ) ;
POLYPRISM  ARM2 HEIGHT  70 AXIS Y                0      0      160      5
ARC  TO 160 80 RADIUS  -37.5 TOLERANCE  1  0      85
ARC  TO 0 0 RADIUS  -42.5 TOLERANCE  1 ;
TO RM501_J3  ADD ARM2  (SHIFT  Y -35 Z -42.5 ) ;
CYLINDER  PIVOT  LENGTH  130 DIAMETER  74 TOLERANCE  1 ;
TO RM501_J4  ADD PIVOT  (ROTATE  X -90 SHIFT  Y -65) ;
CYLINDER  END LENGTH  30 DIAMETER  35 TOLERANCE  1 ;
TO RM501_J5  ADD END  (ROTATE   Y 90 SHIFT  X 37) ;
STOPROBOT  RM.501
```

Figure 3.16: A GRASP Plot of a Mitsubishi RM.501 Robot.

(b) The "Elbow" was the intersection of the centre of the Upper-Arm and the elbow joint Θ_3.

$$Elbow_x = 220.\cos\Theta_1.\cos\Theta_2$$
$$Elbow_y = 220.\sin\Theta_1.\cos\Theta_2$$
$$Elbow_z = 220.\sin\Theta_2$$

(c) The "Fore-Tip" was the centre of the end of the forearm and centre of joints four and five (Θ_4 and Θ_5).

$$ForeTip_x = Elbow_x + 160.\cos\Theta_1.\cos(\Theta_2+\Theta_3-\pi)$$
$$ForeTip_y = Elbow_y + 160.\sin\Theta_1.\cos(\Theta_2+\Theta_3-\pi)$$
$$ForeTip_z = Elbow_z + 160.\sin(\Theta_2+\Theta_3-\pi)$$

These calculations were used for the work described in chapter five for the transformation of obstacles from real space to joint configuration space and for the robot experiments described later in this section. More general details of robot coordinate transforms may be found in Paul(1980) or Craig(1989).

A clockwise turn of the base joint was termed a positive change; as was a move of the shoulder or elbow upwards. The ranges of movement for the three main joints were:-

Θ_1 = -150 to +150 degrees.

Θ_2 = -30 to +110 degrees.

Θ_3 = +90to +180 degrees.

A park configuration was defined for the robot with the joints at:-

Θ_1 = 0 degrees.

Θ_2 = 0 degrees.

Θ_3 = +180 degrees.

In this position the arm was horizontal to the work surface as shown in figure 3.16. The joints were calibrated before use by driving each of the joints slowly to a set of limit switches. The limit switches were read via a simple interface card within the computer. The calibration movement was carefully selected to withdraw the arm keeping the ForeTip at the same height. This prevented damage to the camera and robot in the event of failure in an unusual or dangerous position. Once the limit switch for the shoulder joint was detected the Elbow was raised and finally the base was driven clockwise to the end stop.

The RM-501 had several unusual design features. The two most notable were the mounting of the heavy motors within a tail some distance from the base and a spring mechanism to offset gravity loading on the shoulder joint. These features made the dynamic equations unusual. The robot arm tended to balance the weight of the motors with the arm above the park configuration. In the parked configuration with no drive on the motors, the shoulder joint moved under the action of the spring mechanism to lift the robot arm. These features and their effects on the robot dynamics are discussed in more detail in chapter eight.

Many work-piece handling tasks successfully "robotised" to date are generally suited to first generation robots working in isolation from the rest of the production line. Many of these first generation manipulator operations can be completed by simple point to point motion with relatively uncoordinated joint control for the intermediate motions. These robot tasks will remain economic even as future generations and further developments become commercially available. Because this book is concerned with automatic path planning, the robot position throughout the path had to be more predictable. In most robot applications it is the repeatability which is the important property, but in this case the accuracy of the robot was more important. The robot controller is briefly described in Section 3.8 and some robot properties are discussed below.

Several initial experiments were undertaken to discover the properties of the robot required for the transformations made in chapter five and the path planning work described in chapter six.

Repeatability: The robot was moved to a position so that a pin connected to the end effector made contact with a vertical face. The wooden face was aligned with a line marked on the base surface as shown in figure 3.17.

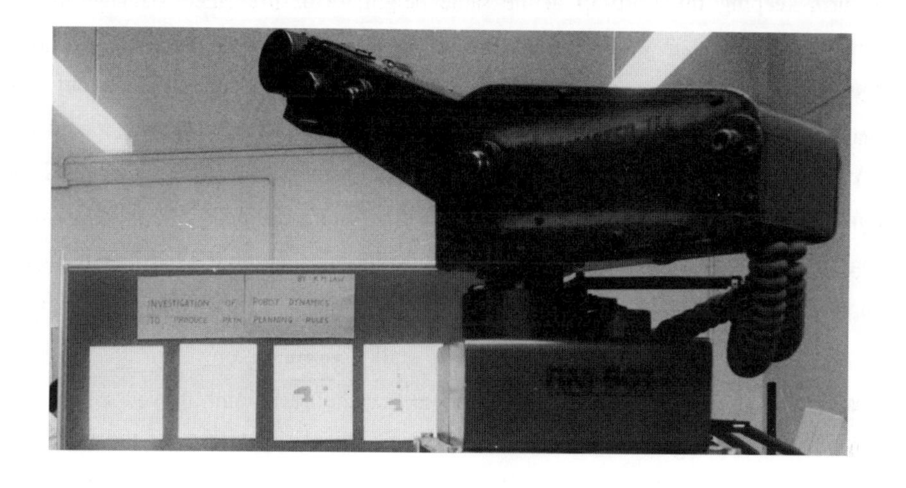

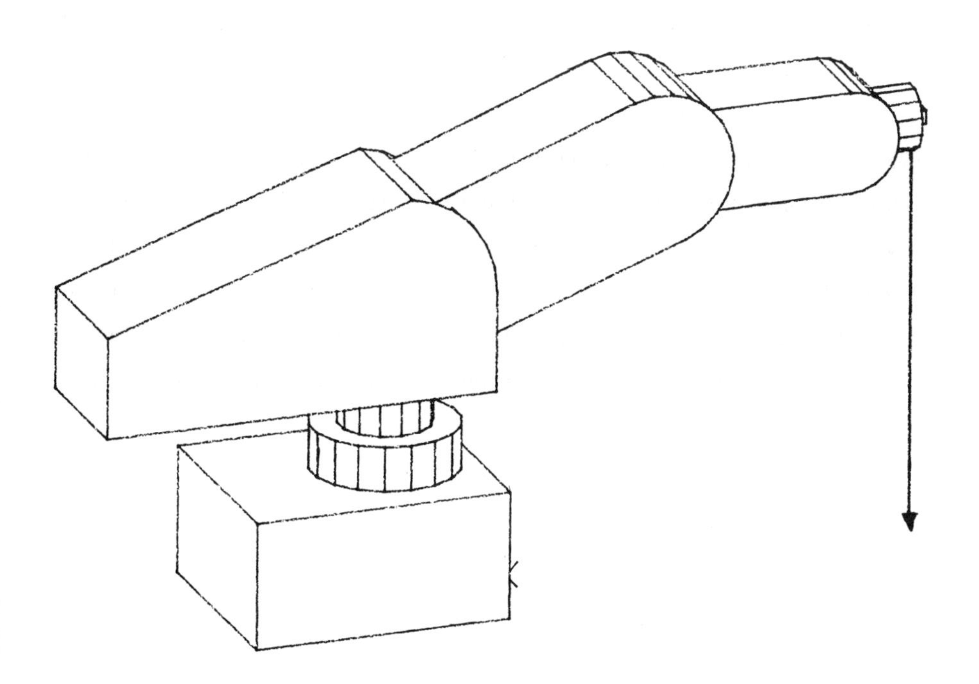

Figure 3.17: A Horizontal Repeatability Experiment.

The robot was programmed to move back and away from the start position to a constant position and then to return. The experiment was repeated with the robot moving to random positions and then returning. The position of the pin was recorded each time against the wooden face and the spread of positions recorded.

A plumb line was connected to the end effector in such a way that the plumb line point just touched the base surface as shown in figure 3.17. The robot was programmed to move up and away from this position and similar experiments were conducted for horizontal repeatability.

The Accuracy: A plumb line was used to measure the accuracy of the position solution presented earlier in the X,Y plane. A rule was used to measure the Z position.

Results: The repeatability experiments were repeated 25 times and the accuracy experiments were repeated for seven different random positions.

Although the repeatability quoted by the manufacturer was ±0.5mm, the experimental repeatability was only within ±2mm when the robot moved away to a constant position and only within ±3mm when the robot moved away to random positions. The difference in these results could be due to the back-lash in the gear mechanisms. The accuracy was within ±3mm .

From the results, the safety margin around the obstacles was set to 10 mm, just over three times the repeatability and accuracy. The safety margin is discussed further in chapter five.

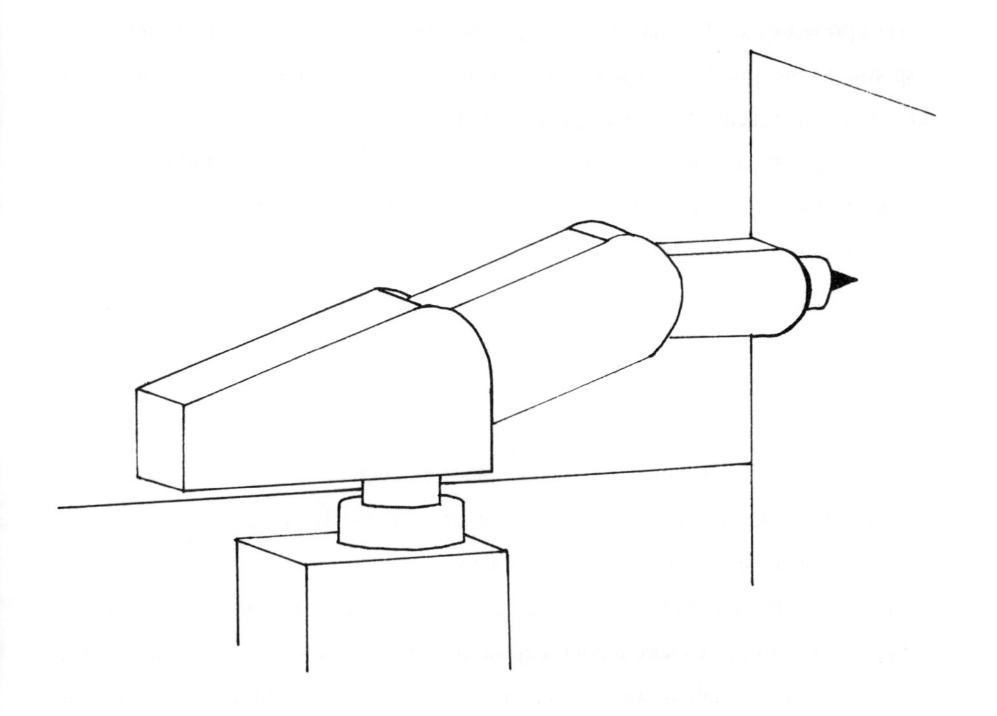

Figure 3.18: The Vertical Accuracy and Repeatability Experiments.

3.8 The Joint Servo Controller.

An analogue and digital loop existed for each joint. The analogue loop was closed around the motor and the digital, sampled data loop was closed via the computer.

The presence of the analogue loop improved the motor time constant and response in the presence of disturbance and parameter variation. This loop used the back EMF from the motor. The digital, sampled data loop was closed via a micro-computer containing the position and velocity control algorithms.

The position feedback was derived from a 16 bit counter clocked by the optical encoders mounted on each joint. This count value had the following relationship with the joint angular positions:

Base = 80 counts per degree.
Shoulder = 83 counts per degree.
Elbow = 80 counts per degree.

The arrangement of the controller is shown in figure 3.20.

The joint servo controller at the Joint control level used look up tables of output voltage values to the D/A converter depending on the position and velocity of each joint. The look up tables were stored in 2-D arrays so that output was a function of the error and the difference of the error with respect to the computer time constant. Output $=f(e, \delta e/T)$.

Maximum drive was output to the joint until a switching region was reached in this joint error space. Through this region the drive was changed to maximum reverse drive until a pure positional control region was reached close to the target position ($e < K$). This is as shown in figure 3.19.

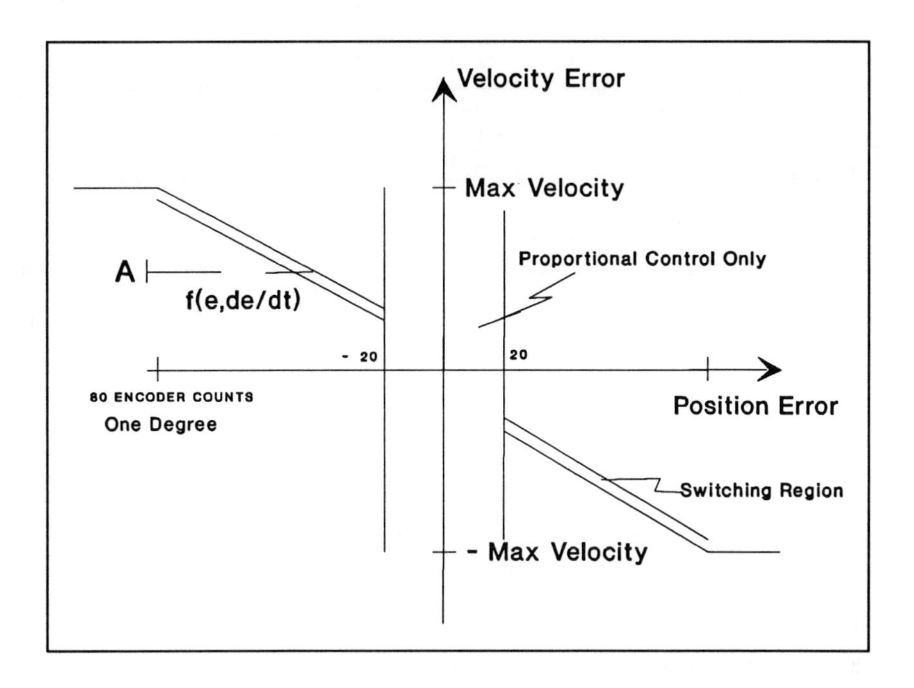

Figure 3.19: Controller Error Signal Phase Plane Description.

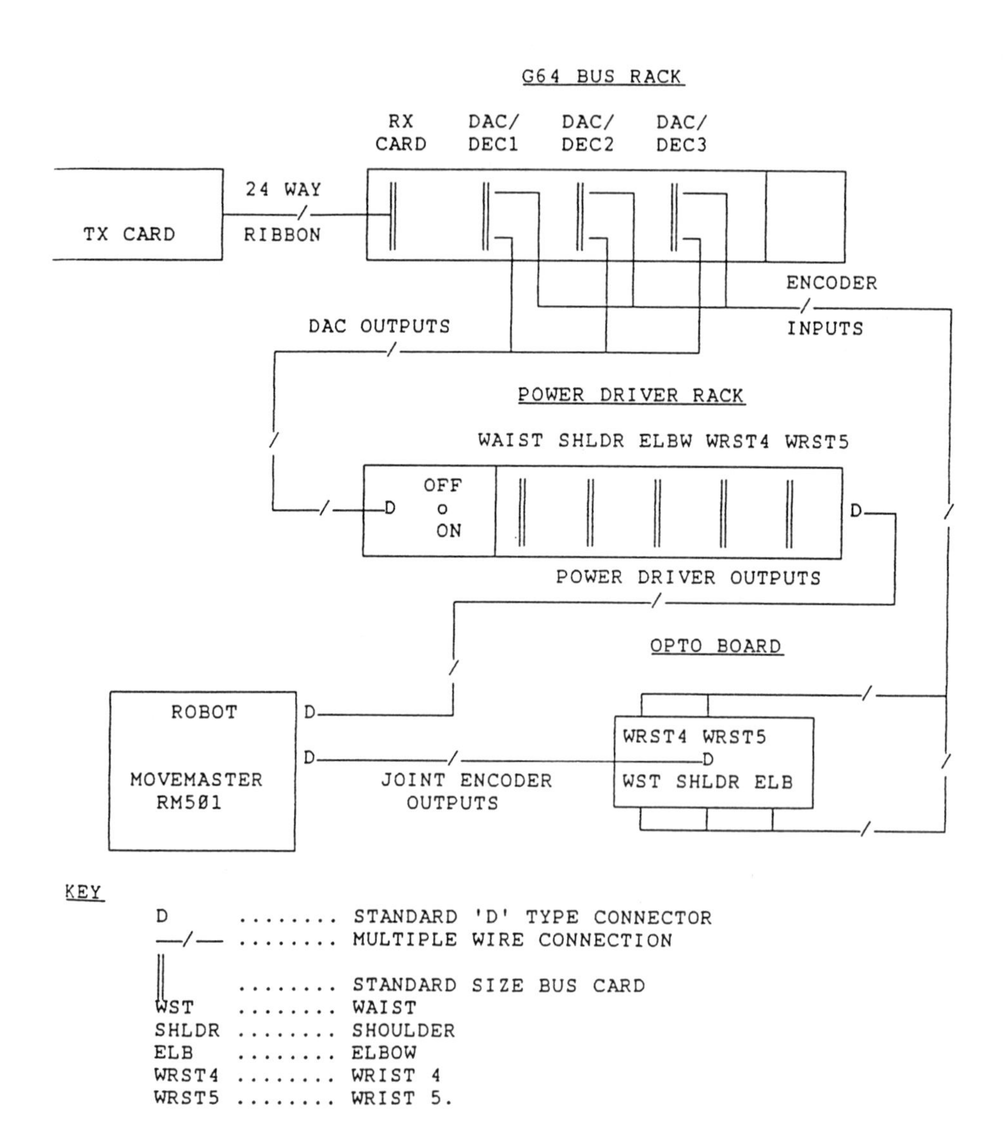

Figure 3.20: A Diagram of typical Device Connections

Chapter Four

MODELLING OF THE ROBOT AND OBSTACLES

4.1 Introduction.

The environment of an industrial robot includes static and dynamic objects. The dynamic environment consists of the robot, objects to be manipulated and obstacles to be avoided. The free space left available to the robot depends on the accuracy of the models used for this changing environment.

As part of his dissertation **Balding(1987)** completed a study of modelling methods and considered the following as important requirements in representing the robot and the work place.

- Fast intersection calculations.
- Ease of use with path planning algorithms.
- Fast model generation.
- Low memory storage requirements.
- Efficiency (in terms of the work-place volume occupied at critical points).

In this chapter, the past work described in chapter two is examined and models are discussed with reference to the above requirements. The models considered can be divided into three categories:-

(a) The Static Environment.

(b) The Robot.

(c) Dynamic Obstacles.

These categories are discussed in the following sections.

4.2 The Static Environment.

Of the modelling methods discussed in Chapter two, the following were investigated for the static environment:

(i) Polyhedral models.

(ii) Constructive solid geometry models.

(iii) Surface models.

Most published computer models of robot surroundings take the form of polyhedral obstacles with flat surfaces and straight edges as this geometry resembles the obstacles commonly found in robot work-cells. These models are difficult to deal with in path-finding calculations and calculation is slow. If both the robot and obstacles are modelled by polyhedral shapes then the accuracy is high but computation time is extended. The GRASP plots shown in figures 3.16 and 3.17 of the last chapter are examples from a system using polyhedral models for all three categories. The system used was the 1990 version, but it could not perform calculations in near real time.

Constructive solid geometry represents conglomerations of objects as ordered binary trees. Figure 4.1 shows an example of a constructive solid geometry tree.

Terminal nodes are either primitive leaves which represent solid primitive shapes, or transformation leaves which contain the defining arguments of rigid motions. Non-terminal nodes represent operators such as rigid motions, intersection, difference or regularized union. In the example, non-terminal nodes are a union (**U**) and a translation. Two solid primitive shapes are shown in cross section in figure 4.1. These are combined using three unions and two translations to form a more complex solid shape. More information about CSG representations may be found in publications by Braid(1973), Braid(1975), and Requicha(1977).

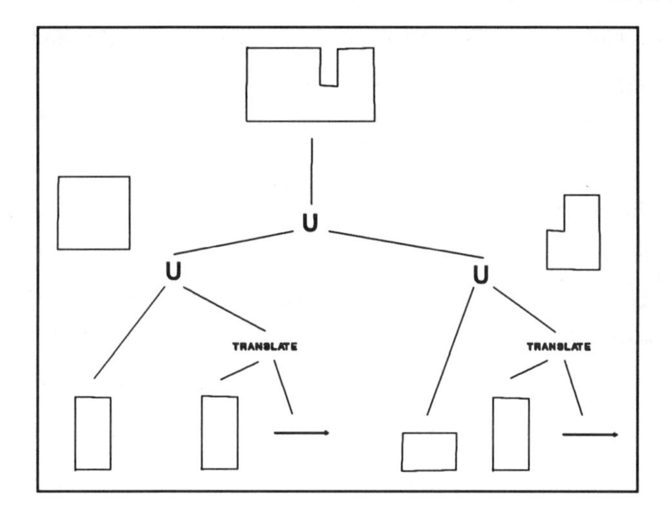

Figure **4.1**: An example of a CSG Tree.

Surface modelling methods have been used to model complex surfaces in detail. An introduction to surface modelling is given by **Ball(1983)**. Surface modellers use complex parametric functions such as Bezier equations to represent the detail of surfaces. These representations are difficult to use for intersection checking as only surfaces are represented. It is difficult to determine whether a point in space is inside an obstacle or not and, consequently, to decide whether surfaces intersect is also difficult.

4.3 The Robot

The requirements for the robot model were similar to those for the dynamic obstacles described in section 4.4. For automatic path planning in real time, the most important factor was the speed of intersection calculation. A constraint was that the robot model selected needed to contain the entire volume of the robot.

A large number of robots have a similar design to that of the Mitsubishi RM 501 robot in that these robots have two major links, (the upper arm and the forearm) and three major joints (Base, Shoulder and Elbow). The simplest possible representation was two lines jointed at one end. Constant distances from the lines were then defined as enclosing the outer casing of the robot. This gave two connected cylinders with hemispherical ends. The advantages of this representation were that the cylinders modelled the robot links efficiently and the intersection calculations between the robot arm and obstacles were simple. The calculations simply consisted of:

(i) In the case of a sphere, finding the distance from the centre of a sphere to the closest point on the line. From this distance was subtracted the radius of the arm and the sphere, to give the distance between the arm surface and the sphere surface.

(ii) In the case of similar 2-D slices, the obstacle model was expanded by the radius of the arm and the calculation reduced to comparing the position of the centre line with the obstacle. This was similar to the 'growing' techniques of **Udupa(1977)**.

The end effector was modelled as a sphere with a radius sufficient to enclose the gripper motors. The work-piece was assumed to be small and enclosed by this sphere. For future work, work-pieces could be modelled easily as additional spheres.

As the path planning algorithm was not designed for angular variation of the end effector, the orientation of the end effector was assumed to be fixed. The end effector position relative to the forearm was defined such that the gripper axis and the forearm were continuous.

4.4 Dynamic Obstacles.

Spheres are the simplest three-dimensional shapes, and Hopcroft et al(1983) described how to calculate intersections among spheres efficiently. These calculations were easily modified to deal with the intersection between lines and cylinders required for the robot model. The method of modelling dynamic obstacles by spheres was initially selected for use in the work described in this book.

Any shape may be modelled by spheres to any accuracy. The greater the accuracy required however, the larger is the number of spheres needed. Experimental work demonstrated that for larger numbers of spheres the computation time increased so that the accuracy of a model was limited by the computation time permitted for the path finding algorithm. This is described further in Section 4.10

In general, it is difficult to decide on the best sizes and positions of spheres to model real obstacles. In practice the number of spheres used to model obstacles were 1, 2 and 4. This made the models simple and speeded up path calculation, requiring little computer storage, while still producing efficient robot paths. When multiple spheres were used for the global path planning method described in Chapter six, there were complications in checking which joint configurations had been checked already for other spheres. This meant there was an increase in processing time, and when multiple spheres were used the sphere model was not the most efficient.

This initial work with sphere models was used to compare intersection calculation speeds for several other models of the dynamic obstacles. Two other models compared favourably:

(a) Similar 2-D slices in joint space.

(b) Six sided parallelepiped.

In all cases it was assumed that the 2-D cross-section of the obstacle in the X-Y plane and the height (Z) of the obstacle were available. This was entirely viewpoint-dependent and could only provide knowledge concerning visible faces and explicit depth information. This was similar to the polyhedral models used by Brooks(1983(b)) in that the obstacles were effectively only two-and-a-half-dimensional. That is, they had a 2-D shape and a height. The 3-D obstacle shapes considered during the work described in this book were:

(i) A cylinder.

(ii) A cube.

(iii) A simple-six-sided polyhedron.

Although the algorithms for sphere calculation were potentially simple, the parallelepiped or similar 2-D planar slices tended to model these 3-D shapes as accurately and in the case of the 2-D slices, more quickly than single or multiple spheres in discretised 3-D space. The method using 2-D slices is described. The models were calculated by considering two pairs of boundaries:

(a) The angles of the base joint, θ_1, which bounded the obstacle ($\theta_1 min$ and $\theta_1 max$).

(b) The maximum distance D_{max} and minimum distance D_{min} from the origin (maximum and minimum radii).

The obstacle was modelled as a series of similar 2-D planar slices. The reference slice was calculated within a boundary of a line from the origin bounded by D_{max} and D_{min} and the limits of the Z axis. The BLOCKED configurations for the shoulder and elbow joints θ_2 and θ_3 were then calculated for this bounded plane and copied for all θ_1 within the two bounding angles, $\theta_1 min$ and $\theta_1 max$.

For the global path planning method described in Chapter six, this reduced the number of time-consuming searches and tests for BLOCKED points that were required. The major part of the algorithm described in Section 4.7 was reduced to copying values within a 3-D graph of configuration space. The obstacle was first modelled as a 2-D rectangle as this was the simplest model which could be derived from the row and column limits of the object under the camera. These limits were derived during the low level image processing described in Chapter five.

The transformation of the models into joint configuration space is described later in this Chapter.

4.5 The Transformation into Joint Space: Introduction.

Once obstacles had been detected and modelled by the methods described in the earlier sections, the data was processed to transform the obstacles into joint configuration space. This was initially achieved for simulated obstacles in the Data Processing level of the main computer and later for real obstacles detected by the vision computer. The programs used are described in Sections 4.6 and 4.7 and in appendix A.

In all cases the robot upper arm and forearm were modelled as their minimum bounding cylinders, with hemispherical ends and the end effector was enclosed by a sphere.

For the global path planning methods it was necessary to transform the obstacles into joint configuration space. A point obstacle in Cartesian space is not transformed into a point in joint space. If the point is within the robot work-space then it is transformed into one or more complex three-dimensional shapes.

Complex shapes may be represented within a computer as geometric shapes, units of space or by approximating the shapes by mathematical curves. The global path planning method represented the obstacles as regions of joint space consisting of small units. The method was not restricted to any particular design of robot and may be used with any number of degrees of freedom. The program presented was based on the implementation for the three major axes of a Mitsubishi RM.501 robot.

For the global path planning method a graph was created which consisted of a three-dimensional structure of unit regions. The 3-D graph had each dimension corresponding to a principal degree of freedom of the robot arm, Θ_1, Θ_2 and Θ_3. The wrist configurations, Θ_4 and Θ_5 were not considered in the graph. Each unit was initially set to 'CLEAR' status and the positions (in joint space) at which the robot intersected obstacles were then calculated. Each unit represented a range of configurations for the robot, in terms of ($\theta_1$cent, $\theta_2$cent, $\theta_3$cent), plus a degree of movement away from these central joint values;

$$\pm\delta\theta_1 \quad \pm \quad \delta\theta_2 \quad \pm \quad \delta\theta_3$$

All units together represented the whole robot work-space and the number of units in the graph, $Node_{Total}$, was given by:

$$Node_{Total} = \frac{(\theta_1 max - \theta_1 min)}{2 \times \delta\theta_1} \times \frac{(\theta_2 max - \theta_2 min)}{2 \times \delta\theta_2} \times \frac{(\theta_3 max - \theta_3 min)}{2 \times \delta\theta_3}$$

where

$\theta_1$max, $\theta_1$min =the upper and lower limits of θ_1.
$\theta_2$max, $\theta_2$min =the upper and lower limits of θ_2
$\theta_3$max, $\theta_3$min =the upper and lower limits of θ_3.

This will later be expressed as:

$$\prod_{j=1}^{3} \frac{\theta_j max - \theta_j min}{2 x \delta\theta_j}$$

If at any configuration in a unit the robot intersected an obstacle, then the unit was set to BLOCKED. If at all configurations in a unit the robot did not intersect an obstacle then the unit remained CLEAR. The path planning problem for the global approach described in Chapter six was then reduced to finding a series of neighbouring units between the START and GOAL configurations that were still CLEAR.

The first method considered for transforming obstacles into joint configuration space was to check each unit of the graph for intersections with each obstacle. This method was slow, taking up to three minutes to calculate for a point obstacle. The program running time was proportional to the number of obstacles and the number of nodes. If the free space was assumed to be larger than the blocked space, a faster method was to consider each obstacle and test for the nodes which could contain the transformed obstacle. This was the method adopted and the algorithm was as follows:

> *For a node in the graph where the robot could intersect the obstacle, recursively test all the neighbouring units to see if they are also within the reach of the robot.*

The programs are described in the following Sections:

4.6 The Transformation into Joint Space: Spheres.

The graph data structure described in Section 4.5 was initialised. The limits of the graph corresponded to the angular limits for the robot's joints within the range of the work-cell and obstacles outside this work-space were ignored. As the graph carried out intersection checks at a limited number of positions, only a limited number of trigonometric solutions were required and these were calculated at the start.

Before the obstacles were calculated all the units in the graph had a flag set to CLEAR status. Four other flags were used with each node; these were:

New obstacle	Forearm tested
Upper arm tested	On list

Each unit code was stored as one byte of computer memory in the array NodeStatus% and the flags used one bit each.

The obstacle data was received from a file or from the vision system and the first task for the program was to read this data.

The task was then split into two sub-tasks, firstly to calculate the upper arm and then to calculate the forearm blocked space on the graph. A configuration was calculated at which the part of the arm under consideration was closest to the obstacle centre. If the forearm was being considered, then the configuration where the foretip was at the centre of the sphere was calculated. For the upper arm, the configuration was calculated for which the centre line of the upper arm pointed at the sphere centre. If the obstacle was within the reach of the link being tested, then this configuration was the first unit for the transformed obstacle.

The base angle was calculated from the X,Y coordinates of the sphere as shown in figure **4.2**. Firstly the modulus (L3) and the angle (Sphθ) from the robot to the centre of the sphere were calculated and a test was conducted to see if the sphere was out of range, in which case no further processing was necessary.

Waist Angle θ_1	=InvTan (Y/X)
Modulus XY	$=\sqrt{(X^2+Y^2)}$
Sphθ	=InvTan (Z / Modulus XY)
L3	$=\sqrt{(X^2+Y^2+Z^2)}$

The cosine rule was used to calculate the shoulder θ_2 and elbow θ_3 angles and this is shown.

L1 =220mm

L2 =160mm

where L1 is the length of the upper-arm and L2 is the length of the forearm.

$$\theta_3 = \text{InvCos}\left[(L1^2 + L2^2 - L3^2) / (2 * L1 * L2) \right]$$

$$\theta_2 = \text{InvCos}\left[(L1^2 + L3^2 - L2^2) / (2 * L1 * L3) \right] + \text{Sph}\theta$$

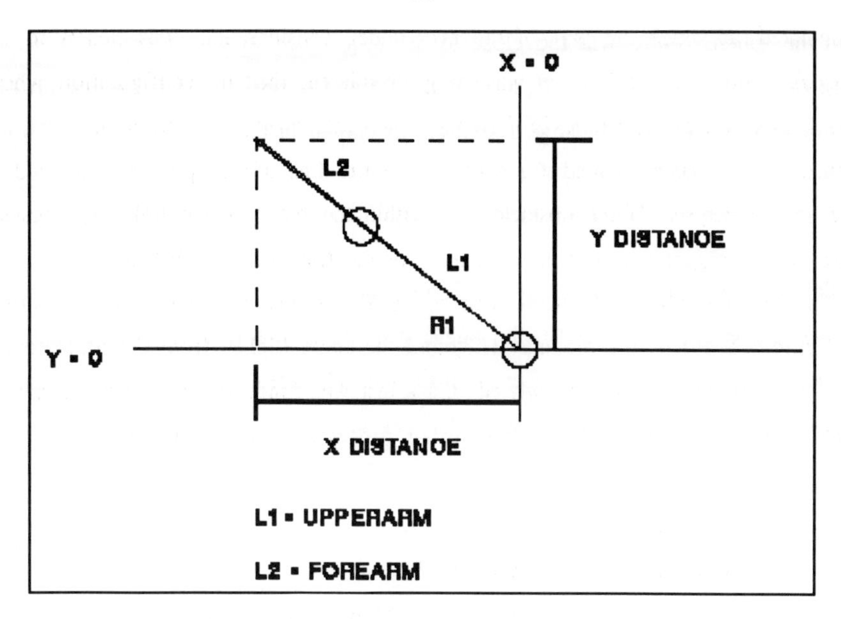

Figure 4.2: Solution for θ_1 (Not to scale).

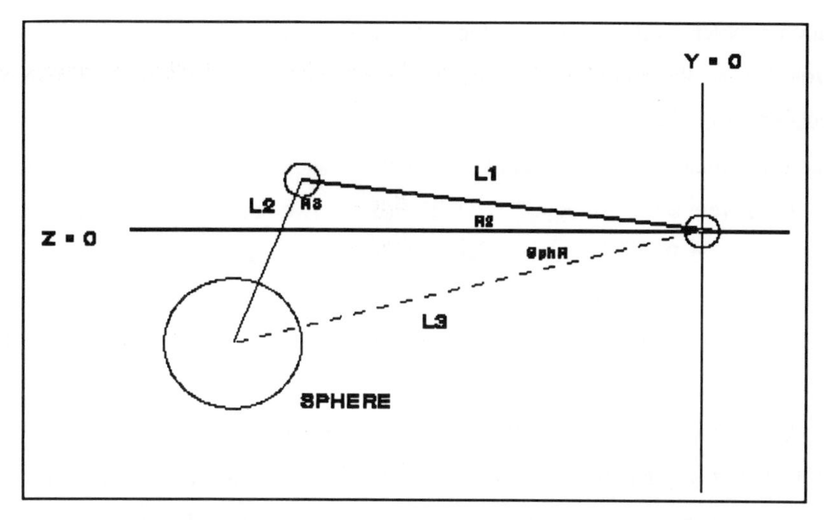

Figure 4.3: Solution for joints θ_2,θ_3 (Not to scale).

If the sphere centre was too close to the robot then θ_3 would exceed its lower limit (θ_3 <90°). In this case θ_3 was set to 90° and θ_2 was calculated using InvTan as the arm formed a right angled triangle as shown in the following code:

```
If θ3 <90° THEN
        θ3 =90°
        θ2 =InvTan ( L2 / L1) +Sphθ
END
```

This gave a starting configuration close to the centre. When the lower limit of θ_2 was exceeded, (θ_2 <-30°), the angle was set to minus 30° and the distance between the upper-arm and sphere centre was calculated (the modulus) using the subroutine FindModulus, from which the cosine could be used to find the new θ_3. The pseudo code for this routine was:

```
If -30° <θ2 THEN
 θ2 =-30°
 Calculate Modulus
 θ3 =InvCos [ ( L1² +L2² - Modulus²) / ( 2 * L1 * Modulus ) ]
END
```

The first configuration was set to BLOCKED. Its neighbouring units were also tested and if they were set to BLOCKED then their neighbours were checked. The position problem was solved using the forward kinematic calculations developed in Section 3.7 and the minimum distance between the obstacle and the robot arm was calculated (provided that it had not completed the calculation before). The method continued recursively until the whole obstacle transformation was found.

All units were set to BLOCKED which had any two opposite neighbouring units which were also BLOCKED. Any units which were on the edge of the now solid obstacle were recorded on a list. All the neighbours of the units on the list were tested, and the process repeated until the surface of the transformed sphere was completely defined.

Nodes which were BLOCKED were stored on a list of units to be expanded later. When a unit was expanded it was retrieved from the list and new BLOCKED points were added to the list. When all the nodes on the list were exhausted the obstacle transformation was complete.

The operation of the lists, the testing routines, the expand routines and fill in routines are described in appendix **B**.

The most important consideration was processing speed. Times for calculating obstacles were recorded during the project and these are presented in the results section of this chapter (Section 4.9).

4.7 The Transformation into Joint Space: 2-D Slices.

The program utilised the data structures described above. These were initialised to form a 3-D graph of joint space and the required trigonometric solutions were calculated at the start. All the units in the graph were set to CLEAR status and similar flags were associated with each node. Obstacle data was simulated or received from the vision system and the first task for the program was to read this data. The two-and-a-half-dimensional model was then created.

Firstly the limits in x were increased by the radius of the upper-arm:

StartRow_Clearance = StartRow - UpperRad%
EndRow_Clearance = EndRow +UpperRad%

The modulus of the ends and centre point on the edge StartCol were calculated with their angles as shown in figure **4.4**. This is shown below for the furthest end from the origin.

Corner(TopLeft, Angle%) =InvTan (StartCol /EndRow_Clearance)
Corner(TopLeft, Modulus%) $=\sqrt{EndRow^2 + StartCol^2}$

The following parameters of the model were found:

The inside radius from the origin. (D_{min})
The outside radius from the origin. (D_{max})
The smallest base angle, ($\theta_1 min$).
The largest base angle, ($\theta_1 max$).

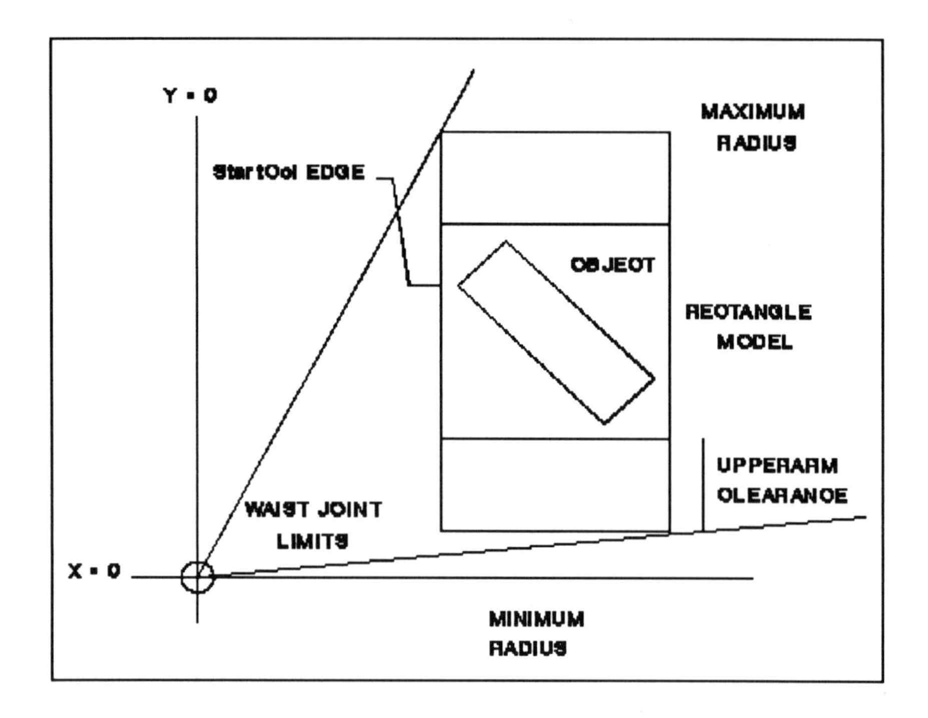

Figure 4.4: Modelling Obstacles using Similar 2-D Slices.

If the obstacle was matched to a template then the height of the obstacle was extracted from the template; otherwise, if the obstacle height was unknown, the height (Z) was set to infinity. The segment was extrapolated to the Y axes so that calculation took place in the Y,Z plane. The modelled obstacle was expanded by the radius of the robot's upper-arm in the Y and Z plane. θ_1 was set to its new lower limit and the inverse kinematic solution was found for all the points within the obstacle, as shown in the following code:

```
FOR Yaxis =(Radius%(min) - UpperRad%) TO (Radius%(max) +UpperRad%)
        FOR Zaxis =-255 TO (Radius%(Z%) +UpperRad%)
                CALL InvKinematics
        NEXT Zaxis
NEXT Yaxis
```

The coordinates in Y and Z were converted to robotic joint angles using the inverse kinematic solution in the subroutine InvKinematics. Firstly the distance from the origin to the Cartesian point (L3) and the angle to the point (Curvϴ) were calculated.

Curvϴ =InvTan Zaxis / Yaxis

sqL3 =$Yaxis^2$ +$Zaxis^2$

L3 =$\sqrt{sqL3}$

The upper-arm was checked against L3 to see if a collision was possible. If within the reach of the upper-arm then θ_2 was set to Curvϴ, and θ_3 was set to BLOCKED between its limits if θ_2 was within its limit.

If L3 was less than the Forearm plus upper-arm then the Forearm collided with the point. θ_2 and θ_3 were calculated using the cosine rule and if θ_2 and θ_3 were within their limits the NodeStatus was set to BLOCKED. The subroutine SetupNodeStatus repeated the NodeStatus settings for θ_1 from $\theta_1 min$ to $\theta_1 max$. This completed the model transformation.

4.8 The Transformation into Joint Space: Other Models.

Although the fastest global transformations were achieved using the solid sphere and 2-D slices, several other models were investigated in order to compare performance. These were:-

(a) Semi Solid Spheres.

(b) Hollow Spheres.

(c) Simple Polyhedral Shapes.

(a) Semi Solid Spheres. The method was similar to that used for the solid sphere except that the expansion routine expanded two nodes at a time from the centre of the sphere. When a CLEAR node was reached, the node was placed onto a spare list (list 3) along with the node it had expanded from. Once all the expansions had revealed CLEAR nodes the subroutine ExpandIn tested the node pairs on list 3 to see if a collision occurred between them. When a collision occurred this was taken to be the edge of the sphere and the NodeStatus was set to collision, otherwise the node was assumed to be on the edge of the sphere. This is shown below.

```
****** Considering the Node Pairs on List3 *****

diffJ1% =(t1% - list3%(NoList3%, RefX%)))
Repeated for t2%, t3%
NoList3% =NoList3% - 1
inc% =1                              ; Initially move the joint +5°.
IF diffJ1% <>0 THEN                  ; If θ1 had been expanded then
  IF diffJ1% >0 THEN inc% =-1        ; if diff is pos then move joint -5°.
   DO
     t1% =t1% +inc%
     CALL Testpos             ; Test the new position.
   LOOP UNTIL (nodecode%(t1%,t2%,t3%) AND 2) =2
  END IF
  IF diffJ2% <> 0 THEN               ; Repeat the process for θ2 and θ3
```

This method became complex when the centre node of the nodes being tested was CLEAR. As the inner node code was set to tested, that point was not retested and expanded. This meant the edge was not clearly defined as the nodes around the inner node were not tested. This led to an attempt to use hollow spheres as described in the next Section.

(b) Hollow Spheres. This program effectively followed the surface of the sphere, setting the surface nodes to BLOCKED so that the path planning program would be unable to enter the sphere. Instead of beginning the process at the sphere centre, the Z coordinate was set to the top of the sphere:

SphereEdge(Z%) =SphereCentre(Z%) - (Forrad% +Radius%)

The program then expanded the nodes as described above, placing BLOCKED nodes onto the list. If more than six collisions occurred in a single expansion then the test node was inside the sphere and the collisions were removed from the list. Where less than six collisions had occurred the robot was following the edge. The CLEAR nodes recorded were assumed to represent the edge of the sphere when passing the data to the path planning computer. The changes made to the expandout routine are shown below:

```
t1% =t1% - 1                    ; Move -5°.
IF t1% >= lowlim%(1) THEN
        CALL Testpos            ; Test the node.
        IF (nodecode%(t1%, t2%, t3%) AND 2) =Collision THEN
        colis% =colis% +1
ELSE
        limits% =limit% +1
END
t1% =t1% +2 * Exptype%                  ; Move +5°.
IF t1% <= highlim%(1) THEN
        CALL Testpos
        IF (nodecode%(e1%, e2%, e3%) AND 2 THEN
                colis% =colis% +1 ; Increment Collision store.
        ELSE
                limit% =limit% +1
```

This was repeated for θ_2 and θ_3. The collision store was checked and if equal to six, then the Foretip was inside the sphere and the collisions recorded from that expansion were removed.

```
misc% =limit% +colis%
IF (misc% =6) AND (TestType% =Foretest%) THEN
        Numonlist1% =Numonlist1% - 6                    ; Wipe off list
```

This method was slow as every time the program entered the sphere, six nodes were tested and removed from the lists. The sphere models were complex in joint space, and following the surface was a complex task (especially when a single sphere could be transformed into two separate shapes in joint space).

(c) Simple Polyhedral Shapes. As discussed in Chapter two, polyhedra are commonly used to model obstacles. The method modelled the obstacles as six sided parallelepiped. The program established the position of the edges of the model in X and Y from simulated data or by calculating the limits of the rows and columns set in the vision program. The height of the object was retrieved from the associated template as described in Chapter five.

In the subroutine TestPos the edge positions were expanded with the model radius of the part of the robot under test (ie upper-arm or forearm), as demonstrated below for an expansion of the forearm in X.

```
Expand_XLow%    =EdgePosition%(LowX%)  - ForRad
Expand_XHigh%   =EdgePosition%(HighX%) +ForRad
```

The Cartesian coordinates of the arm were tested against the expanded polyhedral edge limits, as shown below.

```
IF Expand_Xhigh <= [foretip(X%)] >= Expand_XLow% THEN
        IF Expand_YHigh% <= [foretip(Y%)] >= Expand_YLow% THEN
                IF Expand_ZHigh% <= [foretip(Z%)] >= Expand_ZLow%
                THEN
                        CALL PutonList(t1%, t2%, t3%) ; Node added to list
                        nodecode%(t1%,t2%,t3%)=(nodecode%(t1%,t2%,t3%)OR 2)
                        END
                END
        END
```

4.9 Results.

The most important consideration for the system was that it should be suitable for real-time applications. Times for transforming obstacles were recorded during the project and as an example, the times for the different models to transform the vertical cylinder into joint space are shown below.

The times were recorded with the Z axis of the cylinder at X =0 mm and Y = 310 mm with respect to the origin. The obstacle was simulated.

Model	Time (Seconds)	Number of BLOCKED nodes recorded.
One Sphere	9.8	2455
Two Spheres	15.6	2366
Hollow Sphere	25.8	2476
Simple Polyhedron	29.2	1995
2-D Slices	5.9	2504

Figure 4.5: Table of Transformation times for an Upright Cylinder.

The Sphere Model: An obstacle was modelled first as a single sphere of the smallest radius which would enclose the obstacle. Later, if time allowed it was modelled by two smaller spheres and then four spheres. Nodes set to BLOCKED associated with the first sphere tested usually also collided with other spheres. The forward kinematic solutions did not need to be recalculated for these nodes but the total calculation time increased with the number of spheres because the overhead of calculation for each sphere was greater than the saving in time achieved as the spheres became smaller. This meant the single sphere calculation was faster than the calculations for multiple spheres although the single sphere model was less accurate and had a larger volume.

The problem when using more than one sphere was that the centre of several spheres would be set to BLOCKED (with some surrounding nodes) after the expansion of the first sphere. As these nodes were BLOCKED, later spheres were sometimes not retested so that many nodes were not added to the list.

To overcome this problem the centre was tested for an old collision. If TRUE then the node was placed on a new list, (list 3) where the node was expanded later. The nodes on the new list were then dealt with until the list was empty, meaning that for models using two spheres all the nodes outside the first sphere had been found. The routine is shown below:-

```
IF nodestatus%(t1%, t2%, t3%) =1 THEN        ; Test for Old Collision.
        CALL PutonList3(t1%, t2%, t3%)
                DO
                FOR No% =1 TO numonList3%
                        CALL GetoffList3(t1%, t2%, t3%)
                        CALL ExpandTest
                NEXT No%
        LOOP UNTIL numonList3% =0
END
```

The subroutine Expandtest tested each node after expansion to see if a collision had occurred with the first sphere. If it had then the node was added to list 3 to be expanded at a later date and a bit was set in the flag NodeStatus so that the node was not retested. If the node was CLEAR then the edge of the first sphere had been found and the node was tested for collision against the new sphere using a subroutine Testpos.

```
t1% =t1% - 1
IF t1% > lowlim%(1) THEN
 ******     Test for collision with 1st sphere
 ******     if not tested before
 IF (nodestatus%(t1%, t2%, t3%) AND 65) =1
  nodestatus%(t1%, t2%, t3%) = 64    ; Set node to Tested
  CALL PutonList3(t1%, t2%, t3%)
 END

******      If no collision with other spheres
******      test for new sphere
IF (nodestatus%(t1%,t2%,t3%)  OR 1) =0 THEN
        CALL TestPos
END
```

The 2-D Slice Model: The advantage of modelling the obstacle as a series of similar 2-D slices was that once the collision coordinates of θ_2 and θ_3 had been calculated for a particular θ_1 then these collisions could be repeated for the limits of θ_1 which collided with the obstacle. This reduced the main processing task to copying data rather than calculating forward or reverse kinematic solutions.

The representation of obstacles using similar 2-D slices was the fastest to transform into discrete 3-D joint configuration space. The graphical representation of the blocked angles for different obstacles with their different positions are shown in the following pages.

The results are for obstacles modelled as similar 2-D slices so that only the base angle limits are shown. As the BLOCKED nodes were copied between these bounding angles, the BLOCKED nodes are the same for each of the angles. The bounding angles are shown above each chart.

Shown below is the format for pages 101 to 106 which show a representation in joint configuration space for the different orientations of the six obstacles mentioned in Section 4.4. The obstacles were separated for their configurations into the three with the largest X,Y area and the three with a smaller X,Y area.

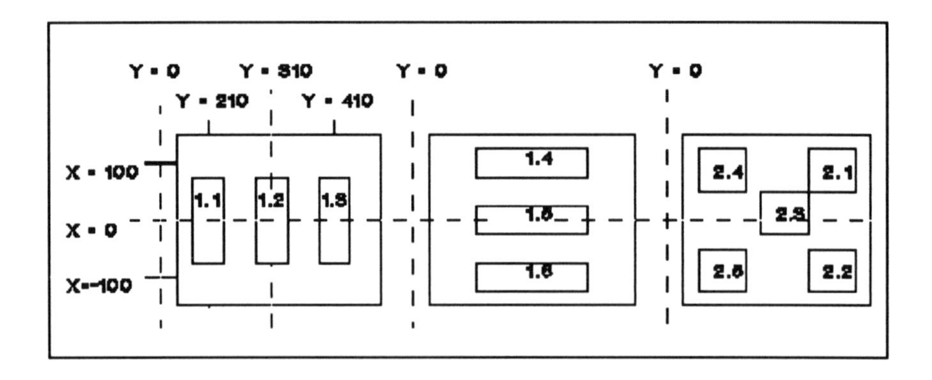

Figure 4.6: The Position of the Obstacles shown in figures 4.7 to 4.39

PAGE FORMAT for the HORIZONTAL CYLINDER. p.92
HORIZONTAL POLYHEDRA. p.93
HORIZONTAL POLYHEDRA. p.94

1.1	1.2
1.3	1.4
1.5	1.6

PAGE FORMAT for the VERTICAL POLYHEDRA. p.95
VERTICAL CYLINDER. p.96
CUBE. p.97

2.1		2.2
	2.4	
2.4		2.5

Horizontal Cylinder.

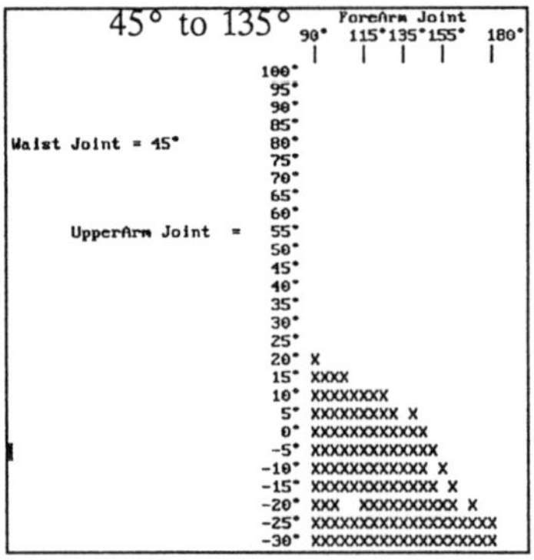

45° to 135°
ForeArm Joint
90° 115°135°155° 180°
Waist Joint = 45°
UpperArm Joint =

Figure 4.7

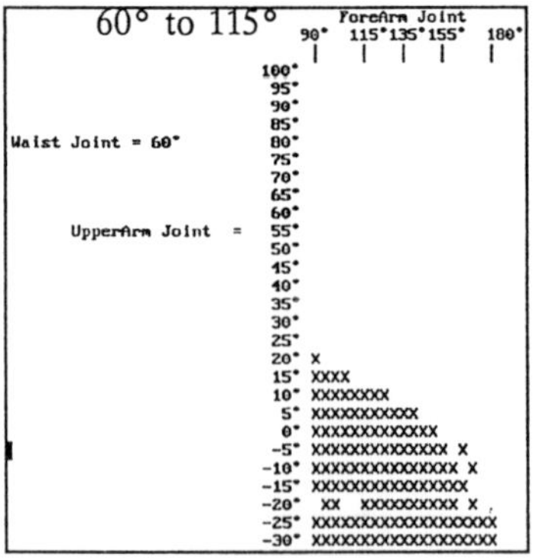

60° to 115°
ForeArm Joint
90° 115°135°155° 180°
Waist Joint = 60°
UpperArm Joint =

Figure 4.8

65° to 115°
ForeArm Joint
90° 115°135°155° 180°
Waist Joint = 65°
UpperArm Joint =

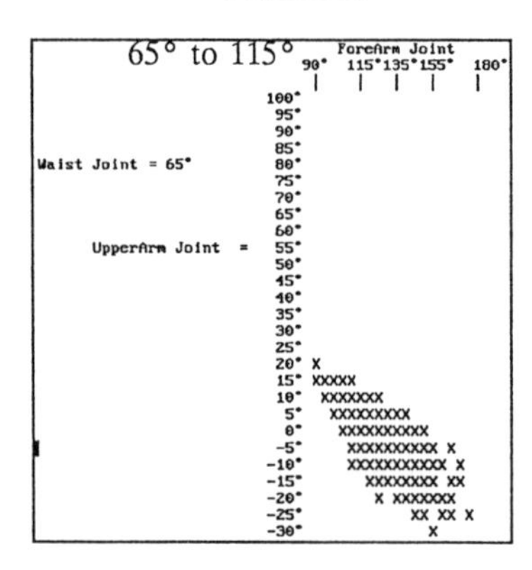

Figure 4.9

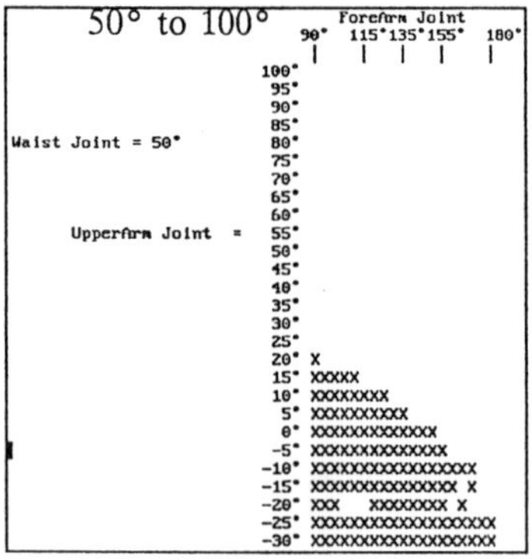

50° to 100°
ForeArm Joint
90° 115°135°155° 180°
Waist Joint = 50°
UpperArm Joint =

Figure 4.10

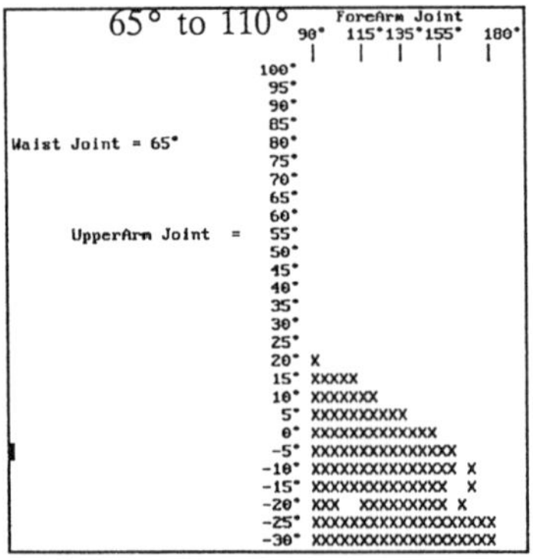

65° to 110°
ForeArm Joint
90° 115°135°155° 180°
Waist Joint = 65°
UpperArm Joint =

Figure 4.11

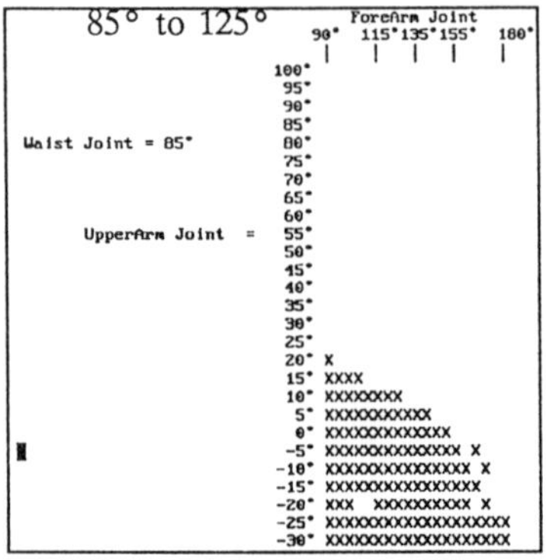

85° to 125°
ForeArm Joint
90° 115°135°155° 180°
Waist Joint = 85°
UpperArm Joint =

Figure 4.12

Horizontal Polyhedra (Maximum Area)

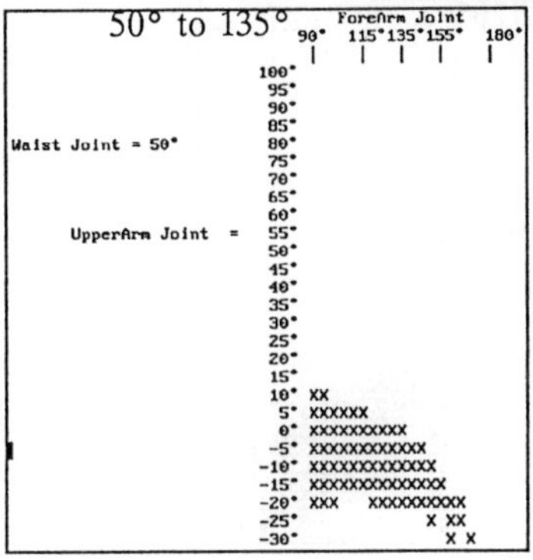

```
50° to 135°        ForeArm Joint
                   90°  115°135°155°  180°
          100°
           95°
           90°
           85°
Waist Joint = 50°  80°
           75°
           70°
           65°
           60°
UpperArm Joint =   55°
           50°
           45°
           40°
           35°
           30°
           25°
           20°
           15°
           10° XX
            5° XXXXXX
            0° XXXXXXXXXX
           -5° XXXXXXXXXXXX
          -10° XXXXXXXXXXXXX
          -15° XXXXXXXXXXXXXX
          -20° XXX   XXXXXXXXXX
          -25°              X XX
          -30°                X X
```

Figure 4.13

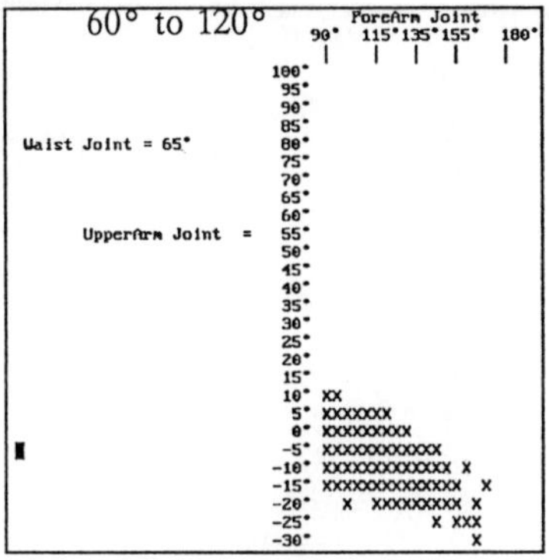

```
60° to 120°        ForeArm Joint
                   90°  115°135°155°  180°
          100°
           95°
           90°
           85°
Waist Joint = 65°  80°
           75°
           70°
           65°
           60°
UpperArm Joint =   55°
           50°
           45°
           40°
           35°
           30°
           25°
           20°
           15°
           10° XX
            5° XXXXXXX
            0° XXXXXXXXX
           -5° XXXXXXXXXXXX
          -10° XXXXXXXXXXXXX X
          -15° XXXXXXXXXXXXXX  X
          -20°   X  XXXXXXXXX X
          -25°             X XXX
          -30°                 X
```

Figure 4.14

```
70° to 110°        ForeArm Joint
                   90°  115°135°155°  180°
          100°
           95°
           90°
           85°
Waist Joint = 70°  80°
           75°
           70°
           65°
           60°
UpperArm Joint =   55°
           50°
           45°
           40°
           35°
           30°
           25°
           20°
           15°
           10°  XX
            5°  XXXX X
            0°   XXXXXXX
           -5°    XXXXXXXXX
          -10°      XXXXXXXX X
          -15°      XXXXXXXXXXX
          -20°        XXXXXXXX
          -25°            XXXXX X
          -30°                XX
```

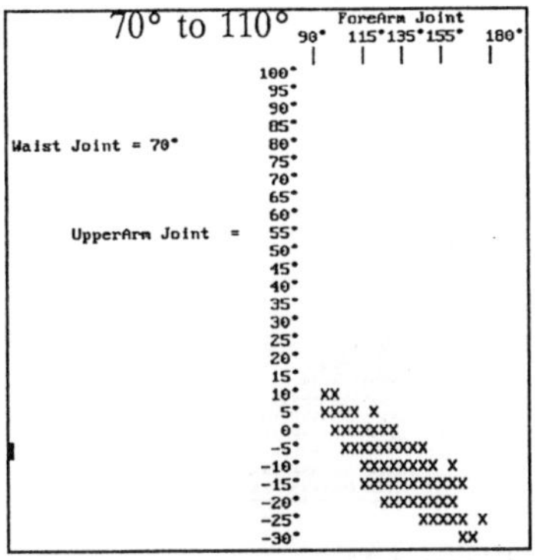

Figure 4.15

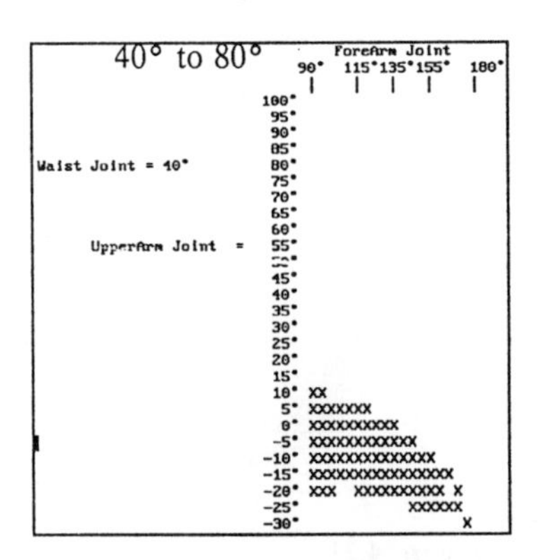

Figure 4.16

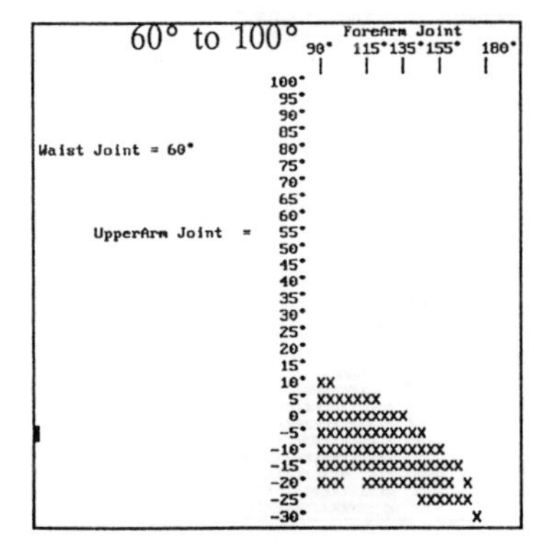

Figure 4.17

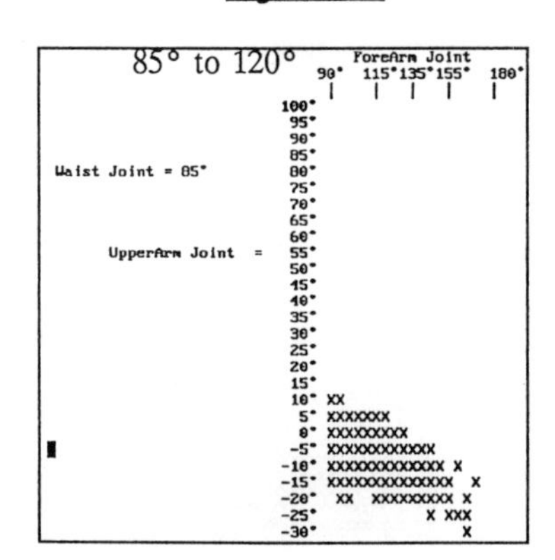

Figure 4.18

Horizontal Polyhedra (Minimum Area)

50° to 135°
ForeArm Joint
Waist Joint = 50°
UpperArm Joint =

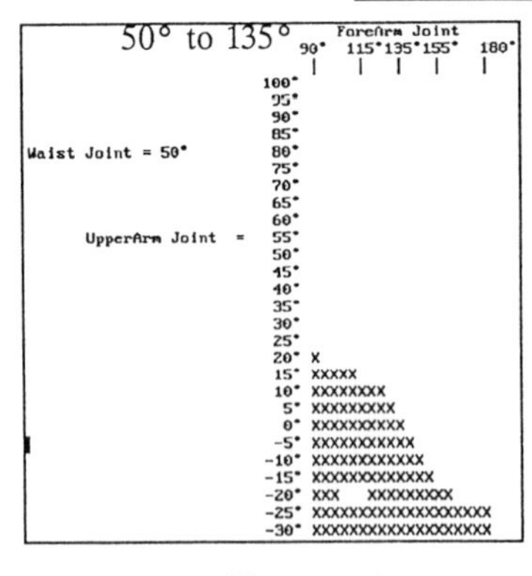

Figure 4.19

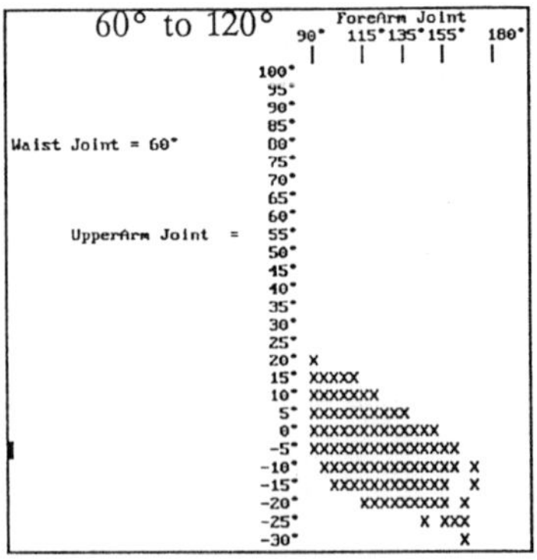

60° to 120°
ForeArm Joint
Waist Joint = 60°
UpperArm Joint =

Figure 4.20

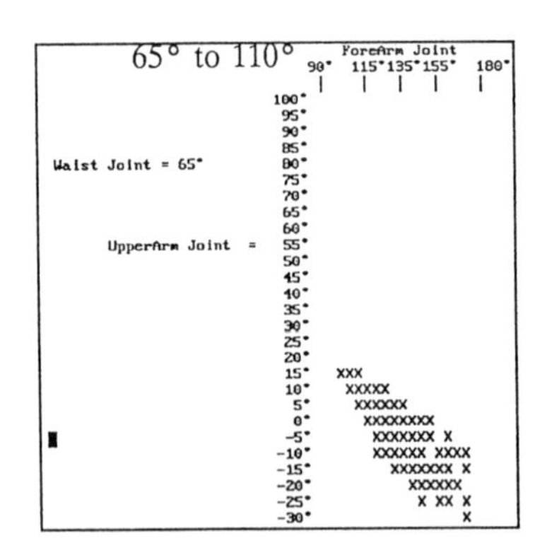

65° to 110°
ForeArm Joint
Waist Joint = 65°
UpperArm Joint =

Figure 4.21

50° to 90°
ForeArm Joint
Waist Joint = 50°
UpperArm Joint =

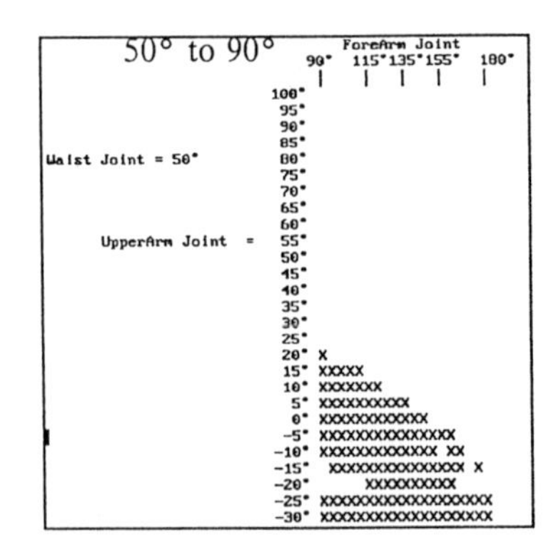

Figure 4.22

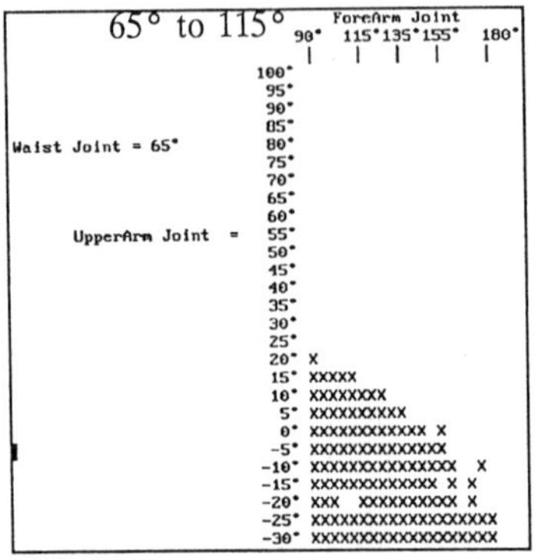

65° to 115°
ForeArm Joint
Waist Joint = 65°
UpperArm Joint =

Figure 4.23

90° to 135°
ForeArm Joint
Waist Joint = 90°
UpperArm Joint =

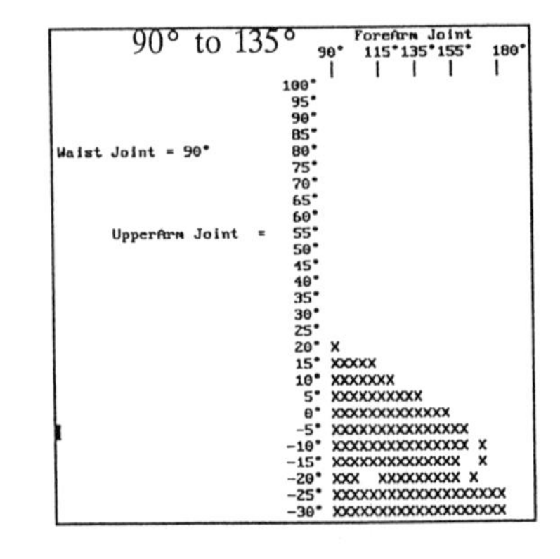

Figure 4.24

Vertical Polyhedra.

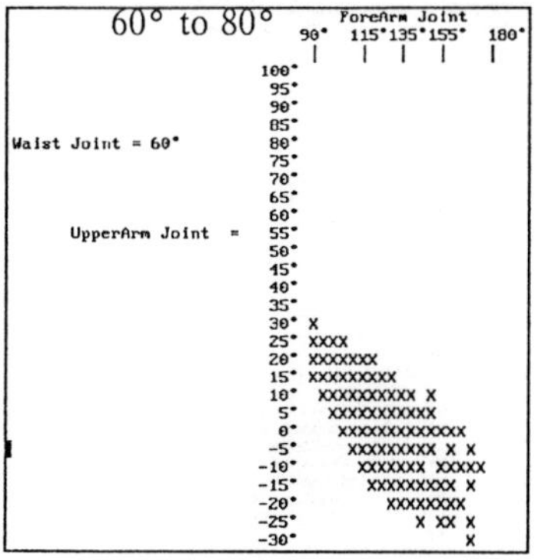

60° to 80°

ForeArm Joint 90° 115° 135° 155° 180°

Waist Joint = 60°

UpperArm Joint =

Figure 4.25

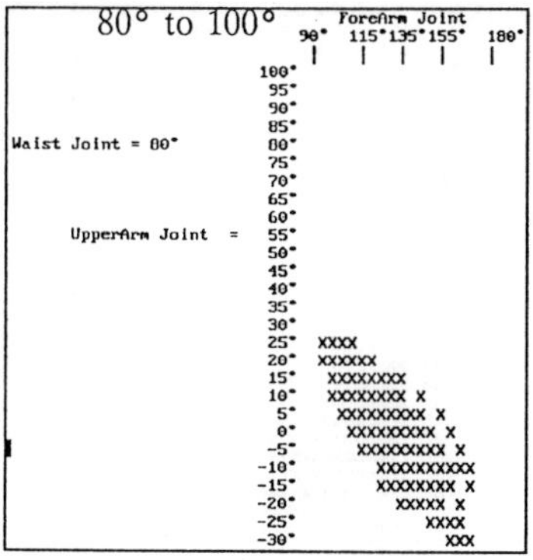

80° to 100°

ForeArm Joint 90° 115° 135° 155° 180°

Waist Joint = 80°

UpperArm Joint =

Figure 4.26

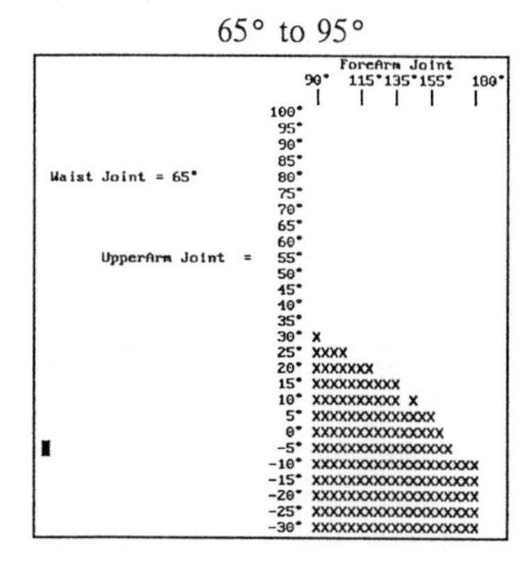

Figure 4.27

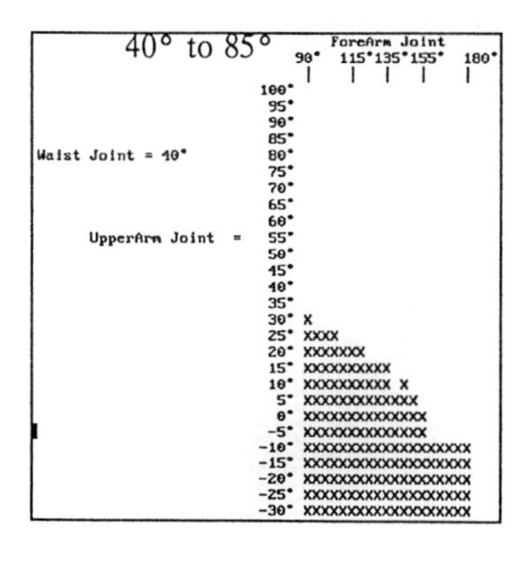

Figure 4.28

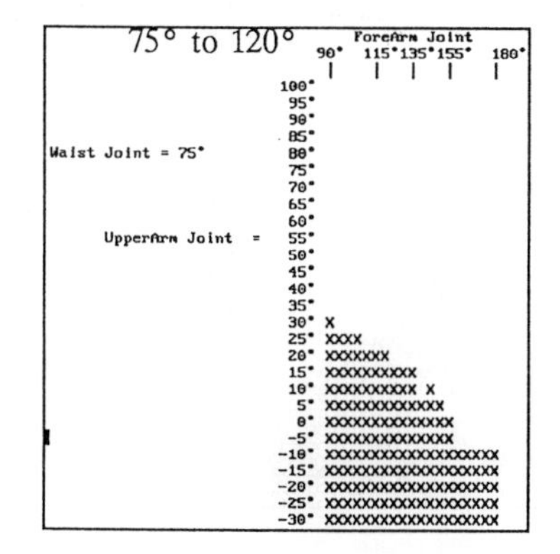

Figure 4.29

Vertical Cylinder.

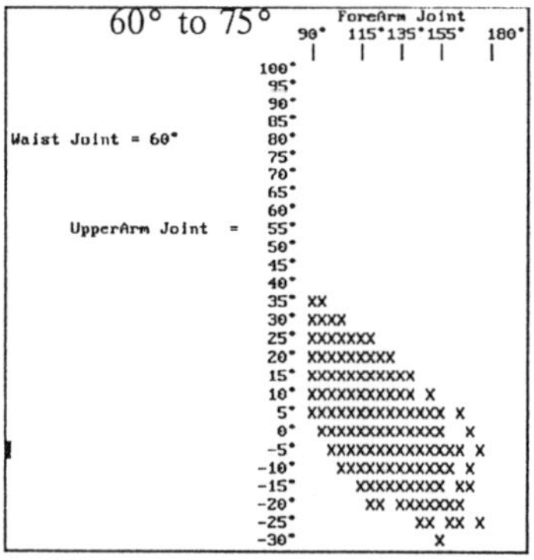

60° to 75°
ForeArm Joint
90° 115° 135° 155° 180°
Waist Joint = 60°
UpperArm Joint =
100° 95° 90° 85° 80° 75° 70° 65° 60° 55° 50° 45° 40° 35° 30° 25° 20° 15° 10° 5° 0° -5° -10° -15° -20° -25° -30°

Figure 4.30

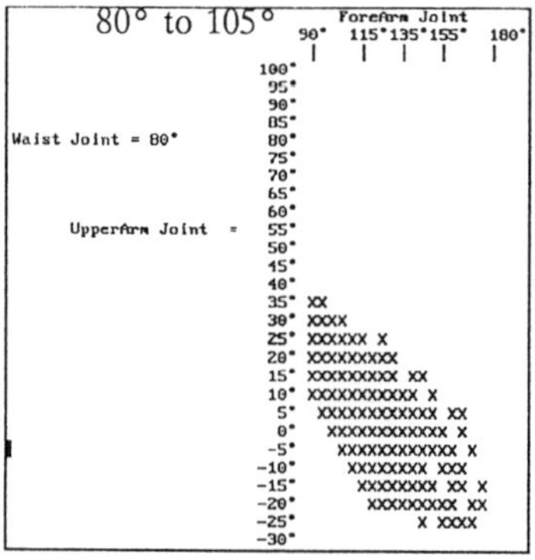

80° to 105°
ForeArm Joint
90° 115° 135° 155° 180°
Waist Joint = 80°
UpperArm Joint =
100° 95° 90° 85° 80° 75° 70° 65° 60° 55° 50° 45° 40° 35° 30° 25° 20° 15° 10° 5° 0° -5° -10° -15° -20° -25° -30°

Figure 4.31

65° to 115°

ForeArm Joint
90° 115° 135° 155° 180°
Waist Joint = 65°
UpperArm Joint =
100° 95° 90° 85° 80° 75° 70° 65° 60° 55° 50° 45° 40° 35° 30° 25° 20° 15° 10° 5° 0° -5° -10° -15° -20° -25° -30°

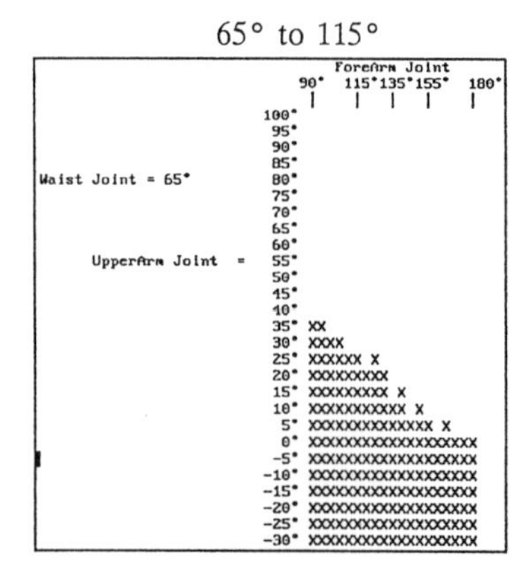

Figure 4.32

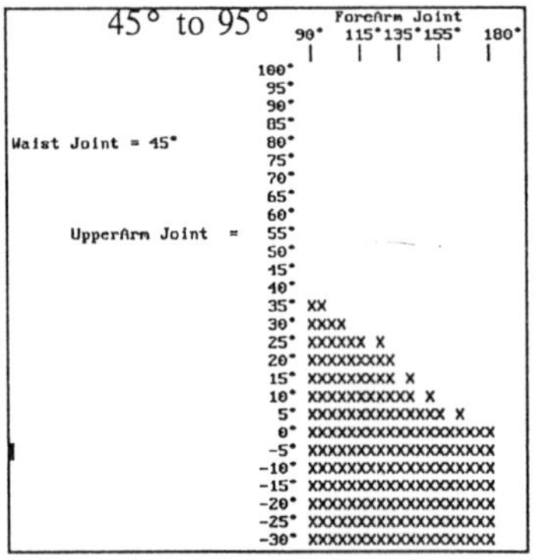

45° to 95°
ForeArm Joint
90° 115° 135° 155° 180°
Waist Joint = 45°
UpperArm Joint =
100° 95° 90° 85° 80° 75° 70° 65° 60° 55° 50° 45° 40° 35° 30° 25° 20° 15° 10° 5° 0° -5° -10° -15° -20° -25° -30°

Figure 4.33

75° to 130°
ForeArm Joint
90° 115° 135° 155° 180°
Waist Joint = 75°
UpperArm Joint =
100° 95° 90° 85° 80° 75° 70° 65° 60° 55° 50° 45° 40° 35° 30° 25° 20° 15° 10° 5° 0° -5° -10° -15° -20° -25° -30°

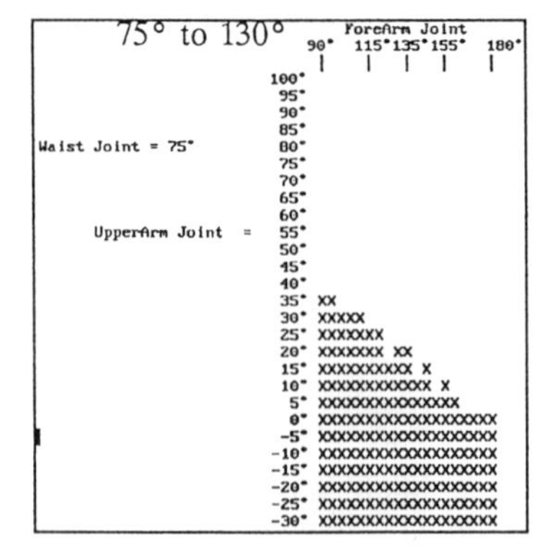

Figure 4.34

Cube.

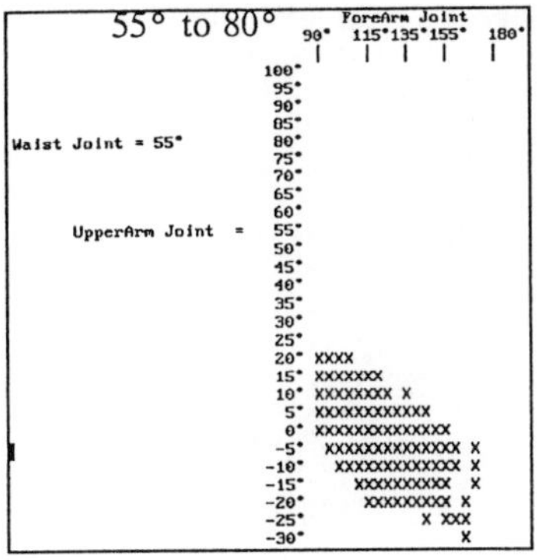

55° to 80°

Figure 4.35

80° to 105°

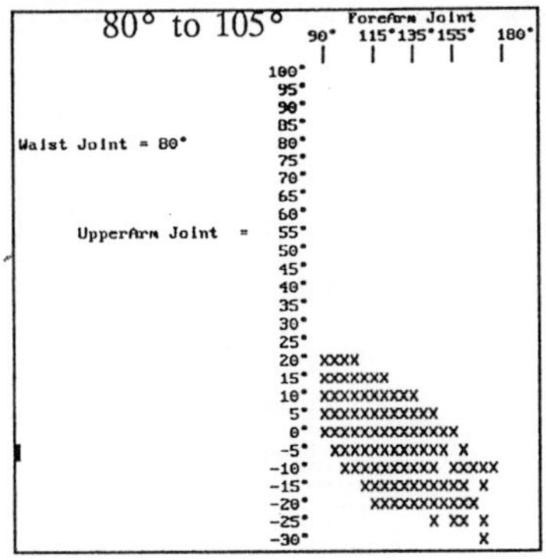

Figure 4.36

60° to 115°

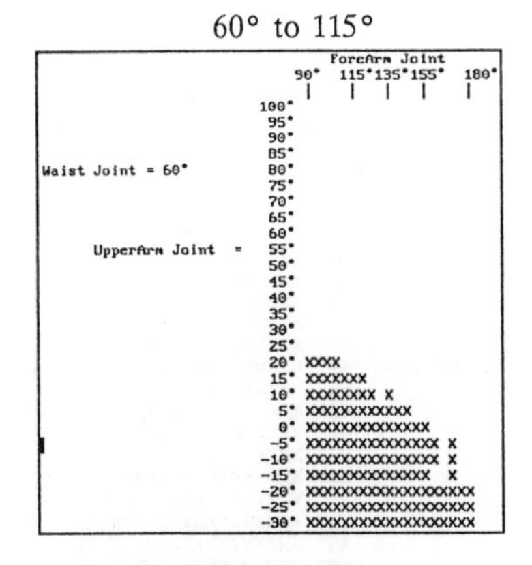

Figure 4.37

40° to 115°

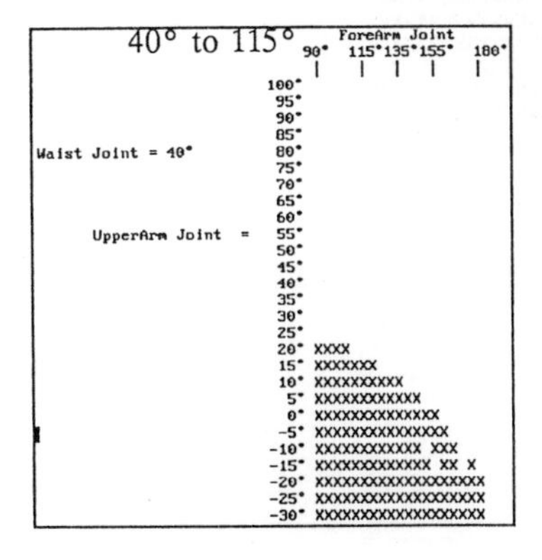

Figure 4.38

70° to 140°

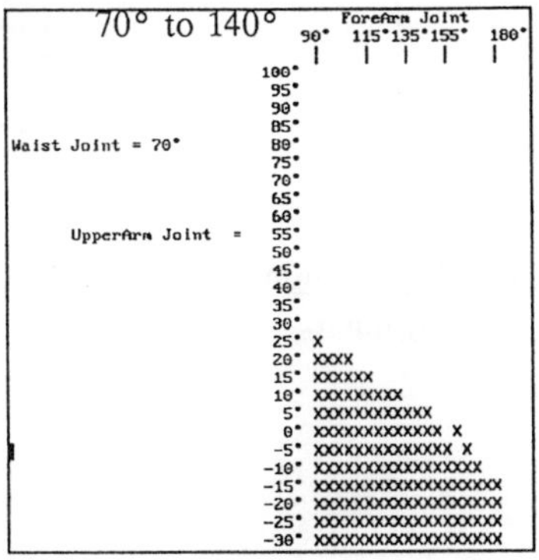

Figure 4.39

UNKNOWN OBSTACLE

60° to 120°

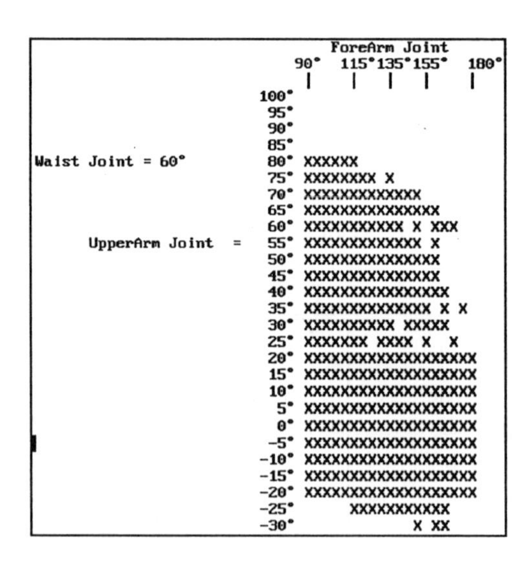

Figure 4.40 Representation in 3-D Joint Spacefor an Unrecognised Obstacle.

The number of BLOCKED nodes for the different obstacles placed in the various positions was as shown below.

Reference Position	Horizontal Cylinder	Horizontal Polyhedra (Minimum Area)	Horizontal Polyhedra (Maximum Area)
1.1	2540	1425	2261
1.2	1890	975	1316
1.3	948	634	786
1.4	1236	616	1072
1.5	1224	608	1064
1.6	1206	628	1058

Figure 4.41: Table of Blocked Configurations.

Reference Position	Vertical Polyhedra	Vertical Cylinder	Cube
2.1	978	854	1106
2.2	986	882	1050
2.3	1238	1364	1456
2.4	1914	1890	2050
2.5	1762	1762	2044

Figure 4.42: Table of Blocked Configurations.

4.10 Discussion and Conclusions.

Once an obstacle increased above a certain size or was moved closer to the origin, part of the obstacle intersected both the Upper Arm and ForeArm joint space. Thus the work-space occupied by the obstacle suddenly increased and the calculation time increased. This can be seen in figures 4.25 - 4.27, 4.30 - 4.32 and 4.35 - 4.37.

For the transformation methods a graph of calculation time vs. discrete work-space volume can be expected to be linear, that is the calculation time for an obstacle was approximately proportional to the number of units tested, the total number of nodes being the work-space volume:

$$\prod_{j=1}^{3} \frac{\theta_j max - \theta_j min}{2 \times \delta\theta_j}$$

The computer time required for obstacle transformations was short. The initial conversion time to model the static environment was slow - up to three minutes of computer time depending on the complexity of the model - but the transformation was only performed when the system was powered up.

The accuracy of the models affected the performance of the Path Planner. High accuracy models required more computation time and therefore longer solution times.

Low accuracy models required links or obstacles to be oversized to eliminate the chance of undetected collisions. Lowering the accuracy led to the rejection of valid solutions.

The world may be modelled accurately by polyhedral, CSG or surface modelling methods but they are complex models requiring complex intersection calculations.

The generation of these models was slow and they would have provided difficult problems for the heuristic algorithms described in Chapter six. For global solutions the transformation of the static environment need only be made once, so that computation time is not a problem. The infrequent initialisation hypothesis of Udupa(1977) also suggested time need not be considered in transforming the static environment. An accurate model was therefore selected and polyhedra were used to model the static environment. For later work using a simple static environment the model was reduced to a single polyhedral shape modelling the work surface.

For dynamic models, speed of calculation was important. The simplest possible intersection calculations for the local methods were made using the sphere model. Calculation was reduced to finding the distance from the robot to a point and subtracting the radius of the sphere to give the distance to the surface of the sphere. The initial work used this model.

Modelling with more than one sphere was considered. As the real environment for a robot becomes more complex so more spheres are needed for the model. It was considered how increasing the number of spheres might increase the accuracy of the model. The case of modelling a unit cube was investigated by Balding(1987). A cubic number of spheres was used, i.e. 1, 8, 27, 64 etc. The spheres formed a regular pattern and were equal in size. An infinite number of spheres was required to model the cube completely but modelling objects using the same sized spheres was inefficient. For example, in modelling a cube using sixty-four spheres of the same size, eight of the spheres are totally enclosed and might easily be replaced by a single larger sphere without increasing the model volume. This is shown in cross-section in figures 4.43 and 4.44.

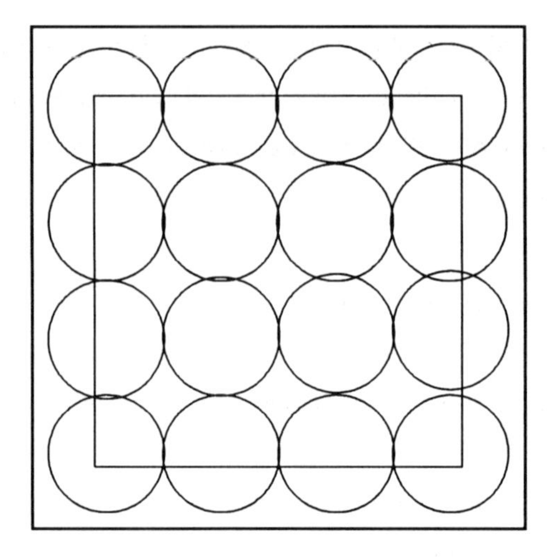

Figure 4.43: Cross-section of a cube modelled by 64 spheres.

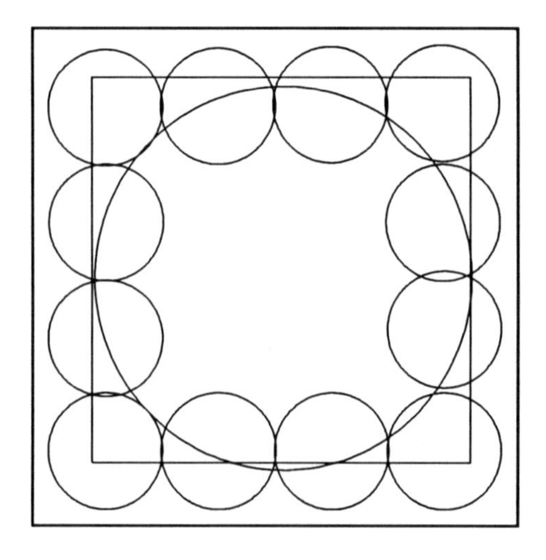

Figure 4.44: Cross-section of a cube modelled using different sized spheres.

To compare the modelling of obstacles using single and multiple spheres, as an example, a model of the cylinder using one and two spheres is compared. The volume of two spheres of radii 35 mm was compared to that of one sphere of 70 mm as shown below.

Volume of Two Spheres

$2 \times 4/3 \times \pi \times 35^3 = 359,188 \text{ mm}^3$

Volume of One Sphere

$4/3 \times \pi \times 70^3 = 1,436,755 \text{ mm}^3$

The area of the two spheres would be much smaller except that the model of the robot must then be considered to find the union volume,

Robot $\cup$ Model

Upper-arm model radius =80 mm

Union radius for a single sphere 70 +80 =150

Union radius for two spheres 35 +80 =115

Union volume of a single sphere $4/3 \times \pi \times 150^3 = \underline{14,137,167} \text{ mm}^3$

Union volume of two spheres $2 \times 4/3 \times \pi \times 115^3 = \underline{12,741,211} \text{ mm}^3$

There was a similar number of collisions for both models. When points within the second sphere were not tested to see if they had collided during the calculations for a previous sphere, this partially explained the lack of improvement in processing time for the model using two spheres.

When the points within a sphere were tested to see if they collided with a previous sphere a saving in processing time could have been expected but in the routine ExpandTest the nodes were continuously expanded until they reached the outer surface of a sphere where tests where conducted to see if the outer surface node collided with another sphere, before the next sphere had been filled. This wasted processing time.

Considering the simple six-sided parallelepiped model, the volume of the model for the horizontal cylinder was less than that of a single sphere or of multiple spheres.

$$\text{Parallelepiped Volume} = (60 + 160)^2 \times (140 + 80) = 10{,}648{,}000\,\text{mm}^3$$

(where 160 was included due to the upper-arm. This was added to allow for the expansion required in X and Y)

This potentially reduced the number of blocked nodes (and therefore the processing time), but the shape and therefore the calculations were more complex so that calculation time increased. This can be seen in figure 4.5.

Using the two-dimensional slice model of the cylinder, θ_2 and θ_3 were only determined for a single slice. This reduced the processing time as this slice of BLOCKED nodes was copied for all θ_1 within the bounding base joint angles. As shown in figure 4.5,the number of BLOCKED nodes produced was similar to other models, so that the intersection volume was approximately the same as for the sphere and polyhedral models. This suggested an equivalent accuracy.

Considering the graphical representation of joint space for the various obstacles shown on pages 92 to 97, for the obstacles with a larger surface area (the horizontal cylinder and six-sided polyhedron), as the obstacle was placed further away from the robot, the range of waist angle collisions reduced. For example for the cylinder:

In figure 4.7 the angular range $=90°$.

In figure 4.9 the angular range $=50°$.

The distance in Y had increased, reducing the total number of BLOCKED nodes shown in figure 4.41 on page 99 from 2540 to 948 and therefore reducing the processing time. This was illustrated when the obstacles were moved from position 1.1 to 1.3. The upper-arm collided with the obstacles at position 1.1 while at 1.3 the upper-arm was out of range. As the obstacles were moved from 1.4 to 1.6 the number of blocked nodes remained constant as did the processing time.

Considering the obstacles with a smaller top surface area, when the obstacles were near to the furthest edge of the work-cell (2.1 and 2.2) the waist joint range was small compared to positions 2.4 and 2.5. As part of the obstacle was out of range of the robot, fewer nodes were BLOCKED. As positions 2.1 and 2.2 were close, the robot upper-arm collided with the obstacle, producing an increase in the number of BLOCKED nodes in figure 4.42 on page 99.

For an unknown obstacle the height was set to the reach of the robot. There were a large number of blocked nodes and the processing time increased but the system could deal with unknown obstacles. An example is shown in figure 4.40 on page 98.

As can be seen from the graphical representation of joint space there were some gaps in the models. This occurred because the inverse kinematic solutions mapped Cartesian coordinates to joint configuration space. As there is no simple continuous relationship between the two spaces a discrete increment of 10mm in

X, Y and Z was used. This meant some nodes were missed in the middle of obstacles. Reducing the incremental value would solve this problem or a dynamic incremental value could be used, so that the value was large close to the centre of the obstacle and smaller once a surface was reached.

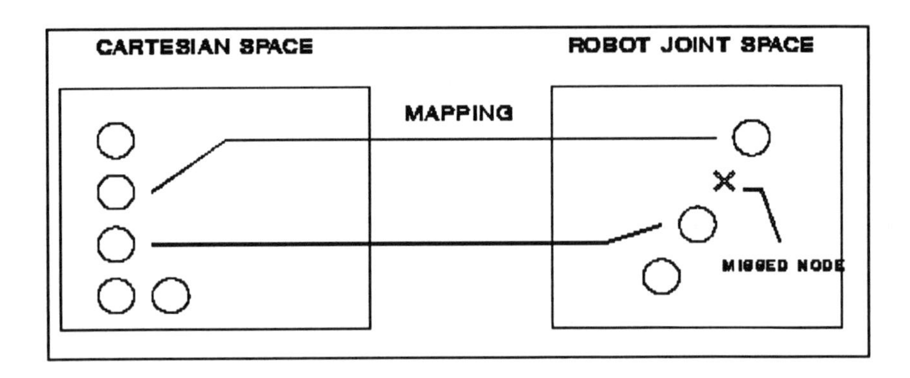

Figure 4.45: Mapping from Cartesian Space to Joint Space for the model using similar 2-D slices.

The heuristic algorithms described in Chapter six were made simpler by the use of spheres, as the distance and direction of the robot to the nearest obstacle were easily calculated. Thus directions could be quickly modified heuristically to avoid collisions.

The detail of dynamic obstacles modelled by spheres was variable, for instance a cubic obstacle away from likely paths could be modelled as a single sphere. Although obstacles were initially modelled as a single sphere as this was fast, for critical items, and if time allowed, the obstacle could be modelled with greater numbers of spheres. The initial work-space volume was reduced by this method, but the critical work-space for path finding was probably not significantly affected.

Of the models considered in this book, the method of modelling obstacles by similar 2-D slices (developed as part of the research work) had the fastest intersection calculation times. This model was adopted for the later work described in this book and will be used in future work. Using the sphere models and 2-D slices described, software models of the dynamic work-place were quickly passed to the main computer by the vision system. The vision system is described in Chapter six and the path planner is described in Chapter seven.

The calculation time for complex obstacles modelled as spheres was short, as each sphere required only four data items (three Cartesian coordinates and the sphere radius). Similar 2-D slices were only slightly more complex, requiring the two bounding angles of the base joint θ_1, the inner and outer radius and a height (five items of data).

For use with the global planning system, the static environment was modelled accurately as several polyhedra, and was transformed into joint space before planning with dynamic obstacles. As this transformation took place once, at the beginning of the program, there was no time constraint. The use of different models for different parts of the work-place is one of the novel concepts presented in this book.

The robot geometric model consisted of two lines connected at the elbow joint surrounded by a skin a constant distance from this skeleton. This model was simple and proved to be fast.

Several different models were considered for the dynamic obstacles and two were selected as they performed the transformation into joint space in the fastest times. The local path planner performed faster when using spheres compared with the other models. It should be noted that there must be a point at which increasing the number of spheres, in order to increase the accuracy of a model, becomes impractical and at this point the Data Processor could change to use a polyhedral model. Although spheres provided the fastest performance for the local path planning algorithms, 2-D slices proved to be faster to transform for the global path planner. This was due to a large amount of the complex processing being replaced by a simpler copying function.

Of the other models considered, none performed favourably with the local path planner but the parallelepiped and the sphere provided favourable results with the global path planner.

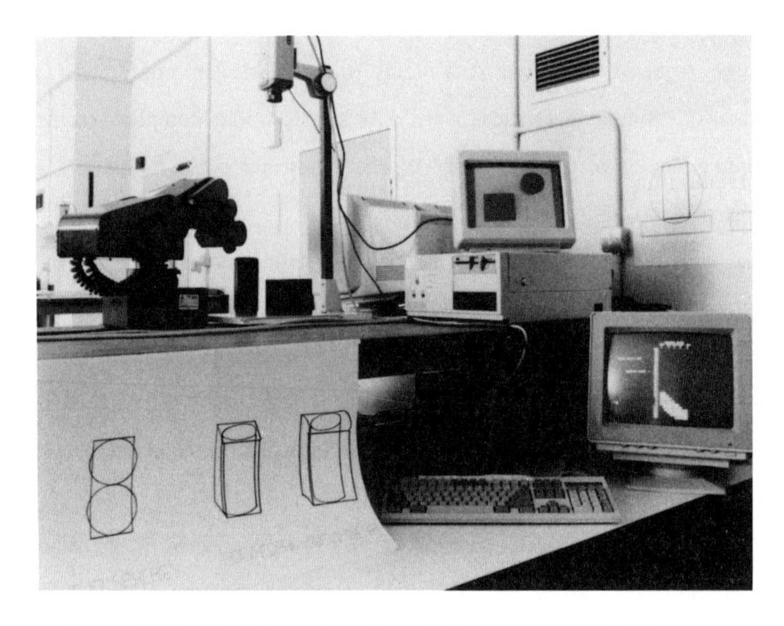

Figure 4.46: Obtaining the plots of the BLOCKED joint space

Chapter Five

SENSOR FUSION AND FORCE SENSING

5.1 Introduction.

Much research effort is aimed towards making robots and complex machines more adaptive and flexible in unstructured or frequently changing environments. This chapter will explore recent methods of sensor fusion and force sensing which can help to achieve this aim.

Research teams are searching for effective and efficient mechanisms for the integration of data from two or more sensors. This data is to be used to create a unified representation of the sensed environment. A prototype sensor system is described and different sensors are integrated on a test rig to investigate sensor fusion.

Conflicts can be detected as soon as more than one sensor is used and the major conflicts tend to be due to differing material types or differing ranges. The conflicts fall into two broad divisions; the same technology providing conflicting information - or - different technologies providing conflicting information. This is dependent on the type and number of sensors in the system. Obstacles detected by the system are assigned certainty levels and placed on a grid which describes the 3-D environment.

The prototype test rig described in this chapter includes a laser ranging system using a Charge Coupled Device camera, an ultrasonic proximity detector and limit switches. Several other sensor inputs are also considered. These include optoelectric, inductive and capacitive sensors.

Earlier chapters have described the desire to create an automatic motion planner in order to increase productivity. This chapter describes some further aims, i.e. of producing systems which may improve the performance of robots and machines.

Performance and speed can be achieved by specifying bigger or more efficient drivers, but this is not an efficient method. An increase in motor torque of 50% can only be expected to give a time reduction of up to 17%, (since time is proportional to the inverse of the square root of torque). To reduce time by a factor of two, the torque must be increased by a factor of four and heat dissipation by a factor of eight. These solutions work by increasing the accelerations so that a stronger and more expensive machinery structure is required. Other methods attempt to improve motions by removing wasted energy.

Existing methods generate paths which may appear simple or obvious to the operator but which may not be efficient for the robot. Once a robot has been programmed to work within a complex system, possibly without the programmer ever seeing the work-place, it may be possible to improve the solution, thus providing the robot with a degree of autonomy. This chapter and Chapter eight explore methods of adapting motions to produce faster and more efficient robot trajectories. To achieve this while machinery is in motion, an indication of the actuator torques is required.

The overseer described inChapter three could be modified to receive information from a Peak Detector. Information on "Vital" and "Non-Vital" movements can be entered by a human operator while programming the machinery movements. "Vital" movements are not changed by automatic adaption algorithms and represent motions which pass close to obstacles or which are specific to some geometry in the work-place; for example, placing a part into a machine. The detection of motor current is described as a means of obtaining torque information.

5.2 Sensor Fusion.

In 1987 Kriegman et al demonstrated that sensor models could be used to map the 'world' using sensor based information. The robot vehicle at Stanford where the work was completed was used as a test bed for a variety of different sensors. The robot had four sensing modalities. The robot received information about its environment from shaft encoders, contact bumpers, ultrasonic sensors and a stereo vision system. The sensors were used to provide a means of navigation through an unknown environment. In this work a means of reducing motion uncertainty due to errors in the sensors was also provided.

In 1986 Ruokagas et al had developed an automation sciences test bed which incorporated vision, acoustic ranging and force-torque sensor subsystems into a hierarchial controlled robot system. Adaptive control based on single, two and three sensors were individually studied using the test bed. More recently in 1987, Luo studied a method of developing an intelligent robot system through the integration of multiple sensors into robot tasks. They argued that multiple sensors generated four different phases of information, that is "Far", "Near", "Touching" and "Manipulation"; these phases are used in the original work discussed in this book. Each phase of information can be integrated into the corresponding phase of robot motions. Most of their efforts were related to data fusion; that is, how to integrate data obtained from two or more sensors into a coherent form.

At the same time Goto and Stentz described a method of integrating sensors in their work on mobile robots. Their system was able to drive a robot vehicle continuously with two sensors on a network of side-walks. Their robot vehicles were equipped with a host of sensors, including colour cameras, a laser range finder and motion sensors such as a gyro and shaft encoders. They identified three types of sensor fusion:

Competitive	- Sensors provide data that either agrees or conflicts.
Complementary	- Sensors provide data of different modalities.
Independent	- A single sensor is used for each task.

In this chapter complementary and competitive fusion types are considered in more detail.

Zheng and Luh categorised multi-sensor integration into data sensor fusion - or - the integration of multiple sensors into an existing control system. Trivedi et al presented a method for determining obstacle positions by integrating vision, range, proximity and touch sensory data. They stated in their research that "External sensory information derived from a variety of sensor modalities is critically important for robots operating in complex, unstructured and dynamic environments". They considered four major issues - sensor modality selection, the low level processing of sensor data, the interpretation of information from multiple sensory domains and decision making with noisy, uncertain or incomplete information. In 1989 Hager and Mintz presented a fusion technique based on a finite element implementation of Bayes theorem. They showed how the method could be extended to cope with uncertain sensor models. The primary limitation of this technique was the heavy reliance on computation. At the same conference Nakamura and Xu presented a fusion technique motivated by the geometry of uncertainties for systems with multiple sensors. The primary aim was to pick up more accurate and less uncertain information by actively utilising redundant information. A general geometrical fusion method was proposed for multi-sensor systems and was also extended to handle partial sensory information.

Crowley described a system for dynamically maintaining a description of the limits to free space for a mobile robot using a belt of ultrasonic range devices. A Kalman filter equation was developed to enable the correspondence of a line segment to the model to be applied as a correction to the estimated position. Two conclusions could be drawn from their system:

(i) An explicit model of uncertainty using covariances and Kalman filtering provided a tool for integrating noisy and imprecise sensor observations into a model of the geometric limits to the free space of a vehicle.

(ii) Such a model provides a technique for a vehicle to maintain an estimate of its position as it travels, even in the case where the environment is unknown.

Grandjean and De Saint Vincent presented a method that fused two kinds of 3-D data. The technique gave a description of indoor scenes with a set of planar faces. They felt that the best approach to tackle the problem of sensor fusion was to use complementary sensors. They addressed the problem of cooperation between a stereo vision system and a laser range finder for 3-D perception. They presented methods for the relative calibration of the transform between the two sensor frames and for 3-D scene modelling using a set of planar faces by fusion of the information provided by the two sensors.

The work described in this chapter seeks to make robots and complex machines more adaptive and flexible in unstructured or frequently changing environments. Effective and efficient mechanisms for the integration of data from two or more sensors into a unified representation of a sensed environment are discussed. A prototype sensor system is described. Different sensors are integrated on a test rig to investigate sensor fusion. Some initial results are presented and the prototype programs are briefly described.

Obstacles detected by the system are assigned certainty levels and placed on a grid which describes the 3-D environment around the test rig.

5.3 A Test Rig to Investigate Sensor Fusion.

A test rig was constructed at Portsmouth Polytechnic to investigate the ideas described in this chapter.

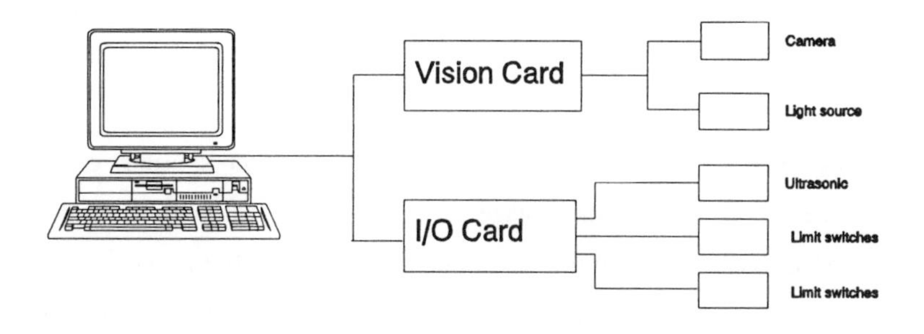

Figure 5.1: A prototype test rig.

The test rig used three different technologies: structured lighting using a laser light source, an ultrasonic proximity switch and limit switches. The test rig was as shown in figure 5.1.

A structured lighting system was created which consisted of a camera interfaced to a micro-computer and a laser light source. Figure 5.2 shows how the range could be obtained from the position of a projected light beam onto an object.

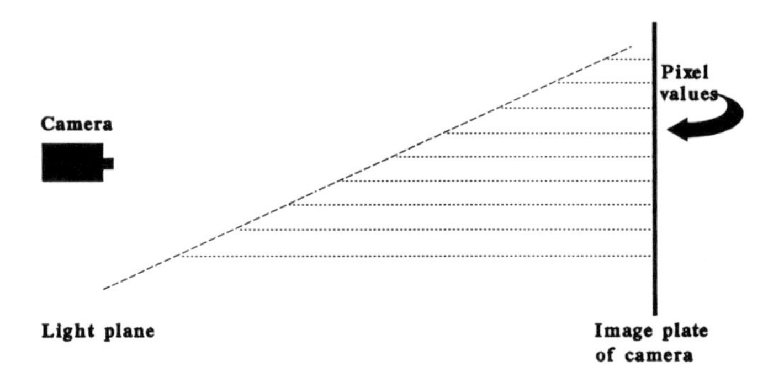

Figure 5.2: Obtaining range information from a laser light source.

In the experimental system the Grey levels of the image were reduced to eight levels before being stored into an array for processing. The stored image was smoothed, a threshold was applied, and the edges of the laser spot could be detected. The image plate of the camera was not linear, so the program calculated the range of an object from a table which gave the range in relation to a pixel value. It can be seen from the graph in figure 5. 3 that this system ranged from 85cm to 175cm.

Smoothing operations were used to reduce noise and other spurious effects that may have been present in the image as a result of sampling, quantisation, transmission or disturbances in the environment. Median filtering was selected, as this method overcomes one of the principal difficulties of neighbourhood averaging in that it blurs the edges and other sharp details. Edge detection involved the computation of a local derivative operator. These operations are described in detail in Chapter six.

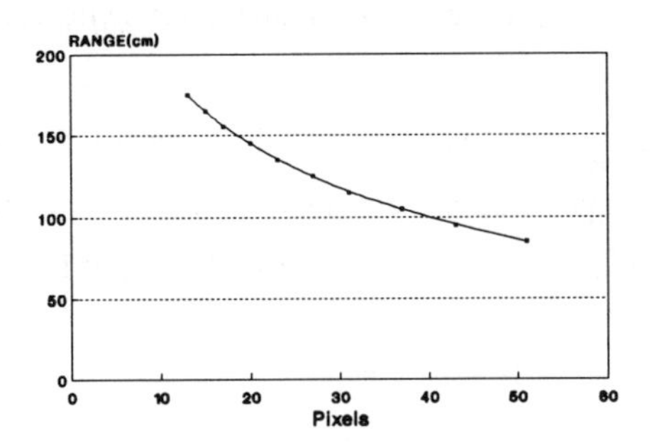

Figure 5.3: Calculating the range of an object.

The magnitude of the first derivative was used to detect the presence of an edge, while the sign of the second derivative determined whether an edge pixel was on the dark or the light side of an edge. The first derivative was obtained by using the magnitude of the gradient at that point, while the second derivative was given by the Laplacian.

For edge detection we were interested in the magnitude of a vector, generally referred to as the gradient and denoted by G[f(x,y)], where:

$$G[f(x,y)] = [G_x^2 + G_y^2]^{1/2}$$

Computation of the gradient was based on obtaining the first order partial derivatives Gx and Gy. The approach adopted was to use first order differences between adjacent pixels. This is discussed further in the next chapter.

The ultrasonic proximity switch consisted of transmitter and receiver circuitry constructed in the workshops at Portsmouth Polytechnic during the writing of the book. When an object was placed in front of the sensor, an output was received from the receiver, telling the system that an obstacle was present. The transmitter circuit consisted of a 555-Timer, a 40KHz transducer and a biasing network. This circuit produced an output as shown in figure 5.4.

The frequency of the circuit was 40KHz. The ultrasound was detected using an ultrasonic receiver crystal. This was prepared during manufacture to resonate strongly to ultra-sound at 40kHz. The electrical output from the crystal was amplified and rectified to produce a potential difference.

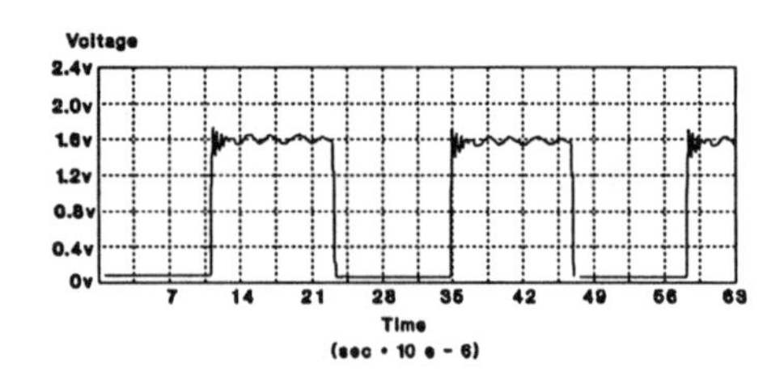

Figure 5.4: The output from the ultrasonic proximity switch.

Finally several limit switches were connected to an I/O card within the computer.

5.4 A Method of Fusing Sensor Information.

A useful method is an extension of the work presented by Abidi in 1991. The algorithm was based on the interaction between the following constraints :

(a) Source corroboration, which tends to maximise the final belief in a given proposition, if either of the sources supports the occurrence of this proposition.

(b) The principle of belief enhancement/withdrawal, which adjusts the belief of one source according to the belief of the second source by maximising the similarity between the two source outputs.

The method was tested by the author using various features from both synthetic and real data of various types. Certainty grids were used to represent the obstacles. This method had been used by Elfes in 1985, who used the certainty grid for off line global path planning. Later that year Moravec and Elfes also described the use of certainty grids for map building. The data fusion method of Abidi is recommended for use in updating the cells within the certainty grid.

In order to create a certainty grid, the environment is divided into elements (denoted as cells). Each cell (i,j) contains a certainty value C(i,j) that indicates the measure of confidence that an obstacle exists within a cell area. The greater C(i,j) the greater the level of confidence that the cell is occupied by an obstacle. If an obstacle produces a reading, the corresponding cell contents C(i,j) are incremented. A solid, motionless object creates a high count in the corresponding cells. Random misreading does not cause high counts in any particular cell. This method provides a reliable obstacle representation in spite of sensor inaccuracies.

The programs written to test the methods combined six different technologies. Simulated technologies included an inductive switch which detected an object within the range 0-15mm. A capacitive proximity switch detected steel within 70mm of the sensor, cardboard within 28mm and glass within 35mm of the sensor. An optical proximity switch detected steel within 520mm and cardboard within 120mm of the sensor.

The test rig included limit switches which detected when the obstacle range was zero. The laser system sensed cardboard or steel obstacles within the range of 400-1500mm. The ultrasonic system sensed steel or glass in the range 400-3000mm.

The program considered the maximum and minimum sensing range for different materials and colours. An error could be introduced on any sensor or sensors for each run of the program. A percentage reliability was generated for each sensor. If the reliability of the sensor was deemed to be less than 65% then the output from the sensor was ignored. The information from the sensors was categorised into the sections suggested by Luo et al; that is:

TOUCHING	- Limit switch
NEAR	- Capacitive
	- Inductive
	- Optical
FAR	- Laser
	- Ultrasonic

This categorisation is used to establish how important it is that any corrective actions are taken.

Problems occur as conflicting information is received from the sensor systems.

These errors can be classified, and once classified, programs can be extended to deal with the conflicts in data. The method extends the certainty grid to include variable flags for each cell which indicate when conflicts occur.

The cells where conflicts are identified are initially just set to BLOCKED. In more complex systems the programs could be modified to learn from these conflicts. Objects which produce conflicts are introduced and the system and the program are informed of the cell position/s occupied by the object. A database is produced containing the learned conflicts and this can be referred to by the computer algorithm when conflicting information is received. Known conflicts can be effectively resolved.

5.5 Typical Conflicts in the Data from Sensors.

Conflicts tend to occur in the following cases.

- When an obstacle is out of range of one sensor and in range of another.
- When an obstacle is made of a material that can be detected by one type of sensor but not by another.
- Due to directivity errors; for example, sound waves may be reflected in a direction which does not return them to an ultrasonic receiver.

Depending on the sensors integrated into a system, different types of conflict can occur: Same technologies giving different opinions - or - Different technologies giving different opinions.

The following list shows a few materials which can provide conflicting information from sensors:

Vision systems	-	Transparent obstacles	Glass.
Ultrasonic sensors	-	Absorbent obstacles	Foam or cardboard.
Inductive sensors	-	Non metallic obstacles.	
Laser systems	-	Coloured obstacles that absorbed the laser light.	

Another aspect of sensor conflict occurs when a sensor detects an obstacle that does not exist or when errors occur within sensors. Effects such as high or low temperature, radiation or water vapour may be likely to trigger these errors.

Comparatively simple programs are able to recognise conflicts that have occurred in the past from a database. A simple decision graph can be applied to the data and from a simple set of rules the conflict can be resolved.

Once a means of determining accurate information from sensors has been developed, the information can be used by other parts of a system to help the machines complete tasks quickly and safely.

Laser systems have limitations, they must be used in a controlled environment. To overcome this, the vision software may be made more complex, but there is still a limit to the improvement that can be made. A 2-D binary camera can be used as these are relatively cheap and levels other than the intensity of the light spot can be set to zero. Methods of obtaining range information from the camera and the laser can be refined to increase the speed of operation. One simple change is to reposition the laser above the camera instead of alongside.

Range recovery through camera motion is being investigated in some research laboratories. This does not require additional hardware, only additional software, but this software is relatively complex.

Programs are able to refer to past conflicts that have occurred by interrogating a database. A simple decision graph and a set of simple rules will resolve these remembered conflicts. Noise in the system can be handled by certainty grids but one set of conflicts cannot be resolved. These conflicts occur due to sensor failure. These failures can be simulated and in the work described in this book, the decision graph attempted to assign remembered results from the database to the conflict. These conflicts are to be investigated further and learning algorithms will need to be extended to deal with these failures.

5.6 Motion Improvement Using Force Information.

The motor drive currents can be monitored to give an indication of the actuator torques. Monitoring the current in a D.C. motor for simple sensing is not in itself a novel idea. Usually it has been used for sensing large forces, although work in 1985 by Naghdy, Sanders and Luk demonstrated its use for sensing smaller forces.

Detailed analysis of actuator current is difficult and high levels of noise are present. This was discussed in Chapter three. The current can be sampled across a small resistance in series with the actuator motor and the signal is passed through a simple filter as described by Sanders in 1987 and mentioned earlier in the book. Current transients may be detected by considering the level and gradient of consecutive samples following new destination signals from the controller. New manipulator destinations may be signalled as actuator moves are generated by the controller.

Once the data from the actuator currents has been analyzed, during the next repetition of the set of movements the joint trajectories may be adapted by changing the controller look-up table. Paths can thus be modified to remove some current peaks in the motor circuits. This can be achieved by :

- Running the actuator motor at low speed instead of stopping at non vital points in trajectories which would normally mean the motor stopped and restarted in the same direction (irrelevant stops). This is signalled by two consecutive current transients in opposite directions as the motor stops and restarts.
- Replacing the look-up table for an irregular stop with a table of a low gain characteristic. This slows the actuator so that non vital destinations are never reached. Velocity is low so that current transients are reduced when the joint stops and then restarts in the opposite direction. This is signalled by two consecutive current transients in the same direction as the motor stops and restarts.

The current waveforms for the two cases are shown in figures 5.5 and 5.6.

A new set of data for the servo motor current is recorded and successive sequences continue to minimise current transients by :

- Increasing the velocity at zero error for irrelevant stops.
- Reducing the gain characteristic for irregular stops.

In the case of a joint restarting in the same direction, (an irrelevant stop), the joint controller changes to a look-up table which slows the joint but which never reaches a zero velocity. A small output may be preset for zero position error. Arrival at a joint via-point is signalled just before it is reached and a new joint target and look-up table is selected as the simulated joint reaches its' joint targets. The change to the current waveform is shown in figure 5.7.

In the case of a joint restarting in the opposite direction, the joint controller switches to a look-up table with a low gain characteristic. Arrival at a joint destination is signalled immediately and the joint moves at a lower velocity, never reaching the via point. A new joint target is selected when the other joints signal their arrival. The change to the current waveform is shown in figure 5.8.

The Controller Software- Various non-linear control algorithms in sub-processes are loaded into the controller. Included is the optimum solution developed for the system by non linearising the experimentally achieved, critically damped control algorithm. These are used to produce look-up tables in memory.

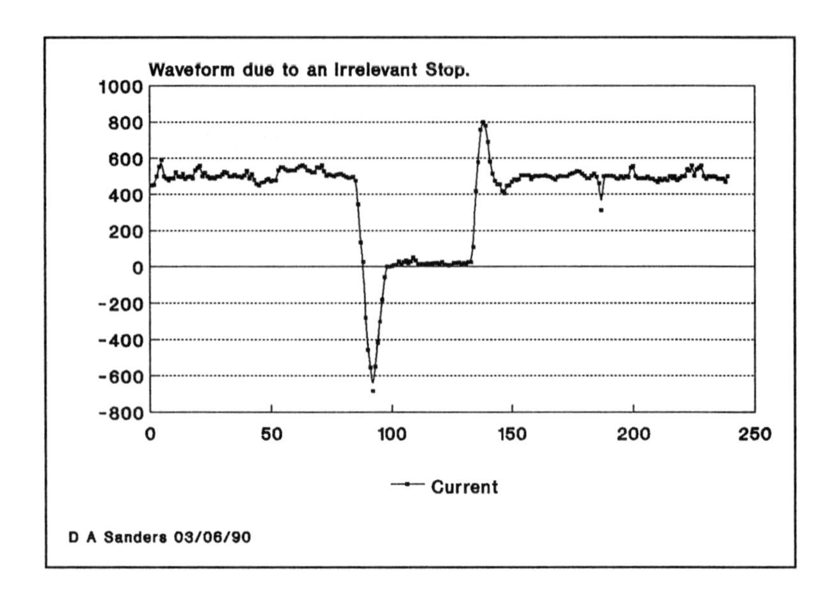

Figure 5.5: Current waveform after an Irrelevent stop.

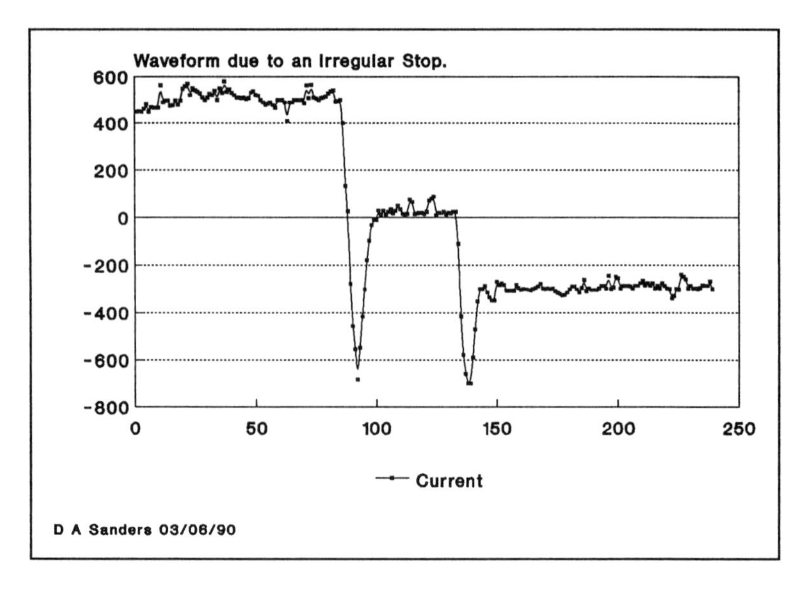

Figure 5.6: Current waveform after an Irregular stop.

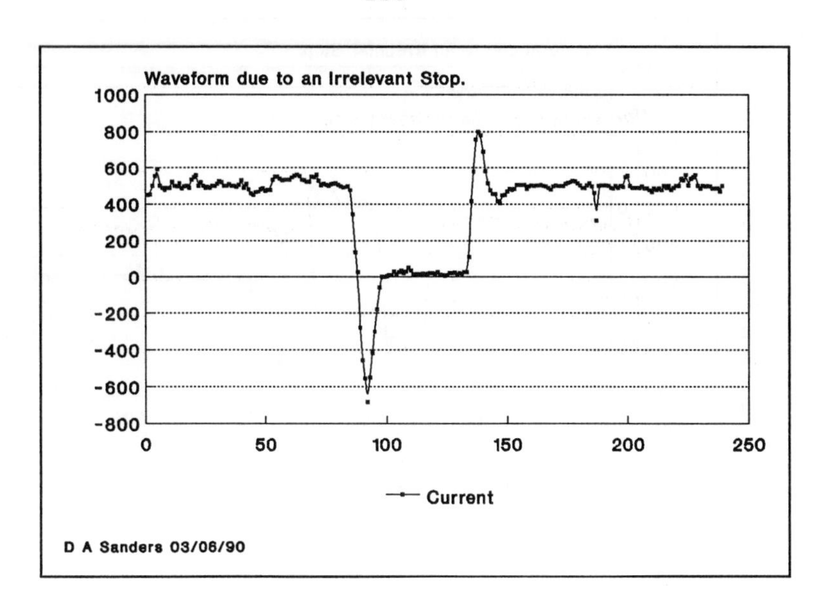

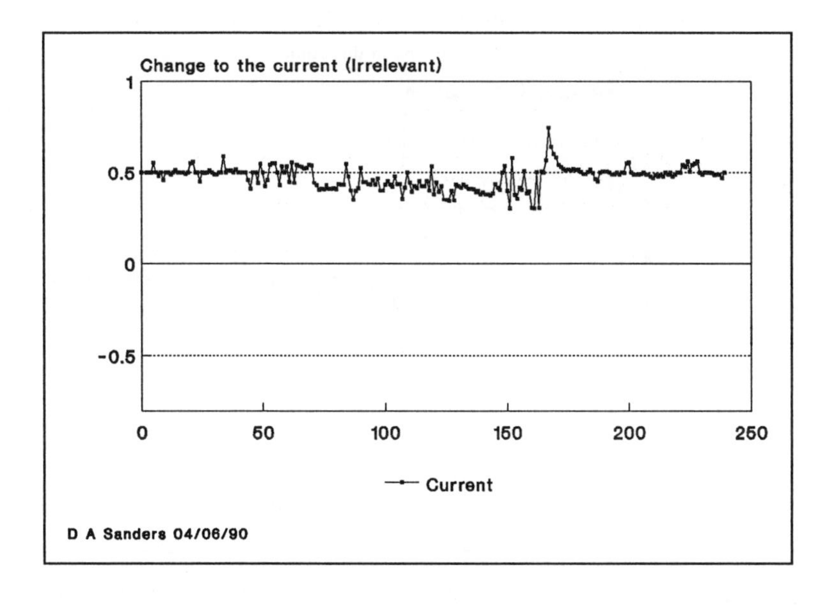

Figure 5.7: The difference in the Elbow Current Waveforms after modifying the trajectory and the path .

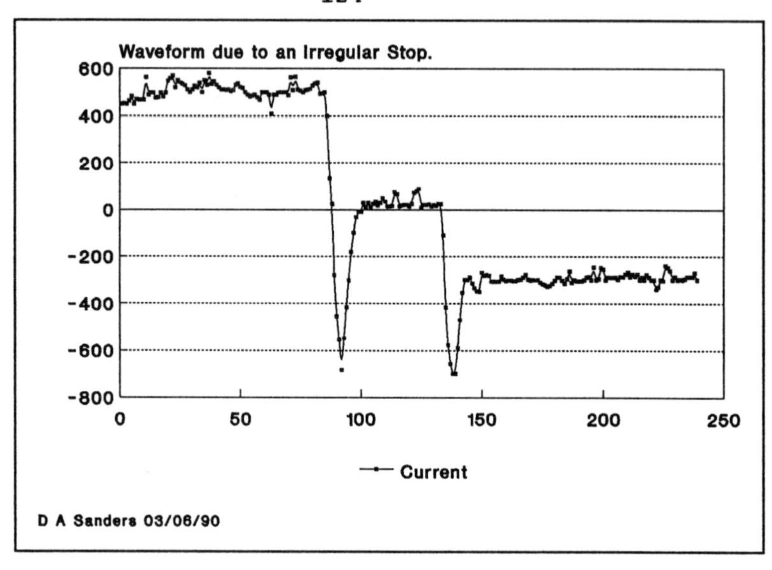

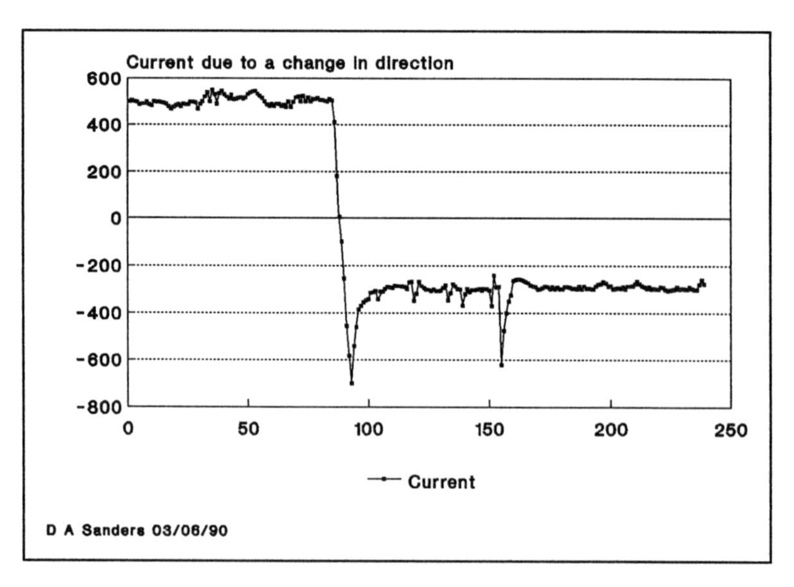

Figure 5.8: The difference in the Elbow Current Waveforms after modifying the trajectory and the path .

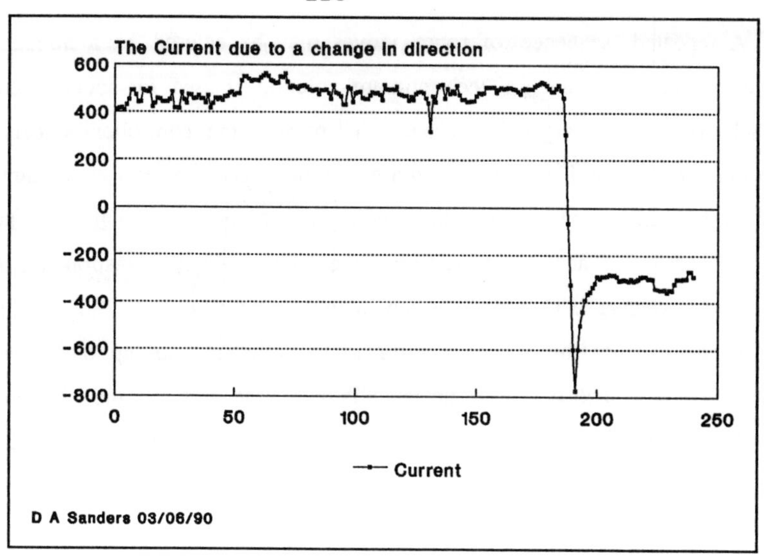

(a)

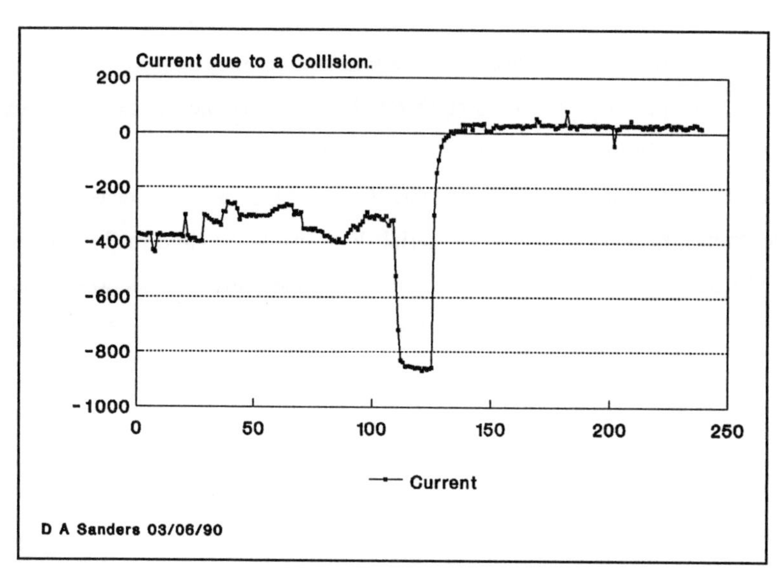

Figure 5.9: The difference between a transient due to a change in direction and a transient due to a collision.

A repeated sequence of robot moves may be entered by a human operator. During the first sequence the optimum solution is used for actuator control. As each joint angle target is reached and passed, the controller signals the main computer. A flag is associated with every move and the least significant bit stores the information concerning vital moves, for example: **1** =vital **0** =non-vital.

If consecutive actuator trajectories are "non-vital", information on the type of non vital change is processed in the main computer and the program is modified in each sequence by passing control to the relevant look-up table for that type modification. The new look up table is then used between the via points.

The Main Computer Software - The main computer samples the D.C current from the servo amplifier driving the actuator. The peak detector is a low level program module in the main computer which collects the information from A/D boards connected to the servo-amplifiers. This information is passed to the path adapter level. The path adapter accepts the information from the peak detector and depending on the type of current peaks, advises the overseer of possible changes to the joint trajectories. This information may be passed to the supervisory level in the controller via a simple serial link. Only the transients associated with non-vital trajectories are considered. If two consecutive transients are non-vital, the relevant change of look up table for the type can be selected. This data can be transmitted via the simple serial link using the information from the flag associated with the move. The data includes the move number as one byte and the type of non-vital transient as one bit in a second byte called a flag. The flag codes are described later in this section.

Once a move has been signalled, transients can be identified in the path adapter by considering the relative level and gradient of four consecutive samples;

$$\mathbf{i}_n, \mathbf{i}_{(n+1)}, \mathbf{i}_{(n+2)}, \mathbf{i}_{(n+3)}$$

Moves are signalled as the controller detects via-points being passed. Each via-point has a number, for example **m**, associated with it and if the wave-form varied monotonically over three consecutive samples then a transient is detected. A forward gradient difference is calculated, so that:

$$\nabla i_n = i_{(n+1)} - i_n$$

where:-

i = instantaneous current.

n = sample number.

∇i_n = nth sample of difference.

then

$$\nabla i_3 = i_{(n+3)} - i_{(n+2)}$$

$$\nabla i_2 = i_{(n+2)} - i_{(n+1)}$$

$$\nabla i_1 = i_{(n+1)} - i_n$$

If the wave-form is varying monotonically, then ∇i_1, ∇i_2 and ∇i_3 are of the same sign. When this is observed, the relative level and gradient can be considered:

$$i_{(n+3)} - i_n = \sum_{n=0}^{n=2} \nabla i_n$$

$$= \pm|\nabla_T|$$

where:

i =instantaneous current.

$|\nabla_T|$ =gradient over 4 samples.

n =sample number.

If $|\nabla_T|$ is greater than a preset limit, for example a constant $|k_g|$, a transient is detected and a transient marker can be set in the main computer. The sign of $|\nabla_T|$ provides the direction of the actuator drive so that for each move, $T_m = +1$ or $T_m = -1$.

If $|\nabla_T| > +k_g$ then $T_m := +1$

If $|\nabla_T| < -k_g$ then $T_m := -1$

If $|\nabla_T| < +k_g$ & $> -k_g$ then $T_m := 0$

where

T_m = Transient marker.

$|\nabla_T|$ = gradient over four samples.

k_g = gradient constant.

m = move number.

Example code in the main computer for detecting a transient is shown.

```
n =n+1
Sample(Gradient())
Gradient(n) =OldCurrent - Current
OldCurrent =Current
Newsign% =SGN(Gradient(n))
IF NewSign% =OldSign% THEN
GradTotal =OldGradient +Gradient(n)
Count% =Count% +1
IF Count% =4 THEN
IF GradTotal >PosTransConst OR GradTotal <NegTransConst THEN
TransMarker%(m) =NewSign%
                Count% =0
                END IF
        END IF
ELSE
        GradTotal =0
        Count% =0
END IF
OldSign% =NewSign%
```

Considering two consecutive transients associated with non-vital moves, the relevant change is signalled to the controller. If noise is introduced into the system and a reading cannot be taken because the signs of the gradient change during the sampling periods, a recalculation can take place in the next pass.

Signals are categorised as "irregular", "unnecessary" or "no-change" depending on the peaks reported by the peak detector. The two lowest bits of the flag are used, for example:-

xx00	=	Non-Vital Irrelevant Move.
xx10	=	Non-Vital Unnecessary Move.
xxx1	=	Vital Move (Not to be changed).

The signal to change the look-up table may be carried in the most significant bit so that an example would be **1xxx xx00**. This would instruct the controller to change to the look-up table for irrelevant moves. The code is shown below:

```
IF (Flag%(m) AND 1) =0
        IF TransMarker%(m-1) =TransMarker%(m) THEN
                Flag%(m) =Flag%(m) +1
        END IF
END IF
```

5.7 Obstacle Detection using Force Sensing.

Problems occur with the improvement methods described in the previous section because similar transients are experienced when a motor is overloaded or when a link meets an obstruction and is forced to stop or slow down. In practice collisions can occur when motions are revised.

Methods of discrimination can be considered.

Forces exerted in Cartesian space can be related to forces in the joint variables by a Jacobian matrix. The calculation of this matrix was described by Orin & Schrader in 1974.

$$|\mathbf{F}_t| = [\mathbf{T}+\mathbf{Q}].\ [\mathbf{J}]^{-1}$$

where: $\mathbf{F}_t$ = vector of Cartesian forces,

$\mathbf{T}$ = vector of joint variables,

$\mathbf{Q}$ = vector of external forces.

These joint forces can be used to detect collisions by monitoring the joint motor currents. Motor current waveforms during contact with hard obstructions have a larger amplitude than transients associated with changes in direction. This is shown in figure 5.9. These collisions are detectable. Contact with softer objects is more difficult to discriminate.

In addition to using the signals to the mixer from the tacho-generator, the system described in Chapter three can be modified to also calculate the velocity in the software from the changes in position and the change in time. This velocity may be used to consider suspected collisions by comparison with the error demand value, e_d. Velocities are monitored and a large error with a low velocity suggests an overload or collision.

The controller informs the overseer in the main computer when a demand signal is generated via a serial interface. Any transients not associated with the generation of new demand signals are regarded as collisions by the overseer.

It is possible to use software calculation with information from joint motor currents for force sensing. The motor current varies in proportion to the motor torque and manipulator forces are transmitted to the joints as the motors attempt to overcome these forces. This generates transients, and current peaks appear on the current wave-forms. The torques can be detected by monitoring the current.

Collision detection: A collision may be notified from three levels.

- In the peak detector within the main computer.
- In the strategic level of the controller.
- In the overseer within the main computer.

- **The peak detector**. The amplitude of the current is monitored and compared to preset limits. Transients exceeding these limits are regarded as collisions and an instruction to stop is passed to the controller.

- **The strategic level**. The error demand value is compared with changes in absolute joint position. A small change in position associated with a large error demand is assumed to be a collision.

- **The overseer**. Unexpected transients received by the peak detector not associated with a marker from the supervisor in the controller are regarded as collisions.

Current peaks are detected by considering the level and gradient of consecutive samples following a new destination marker signal. Actuator moves are signalled to the overseer by the supervisory level of the controller as new manipulator destinations are generated. This allows the detection of unusual current peaks. Transients not associated with a new destination signal are assumed as collisions.

The velocity of the joint is calculated in the controller from the monitored absolute positions. Any velocities approaching zero are compared to the demand error signal.

The sampled data are analyzed and information on adaptable trajectories is considered in the main computer. The joint trajectories are adapted by changing the controller look-up table in the robot controller.

Revised Software for the Controller. The programs work as described in the earlier section except that:

- The velocity is calculated and compared to the demand signal.
- A sub-routine is included to stop the robot if a collision is detected.

After each position reading is taken, the timer is interrogated, and providing the timer has not changed in excess of a preset limit, the velocity is calculated. If three consecutive velocity readings are low and a large error demand exists, a collision is assumed and the manipulator is stopped. An example of typical code is shown.

```
TempTime =TIMER - OldTime
IF TempTime <TimeLimit%
        Vel =(NewPos - OldPos) / TempTime
        If Vel <VelLimit% AND Error >ErrLimit% THEN
                ZeroJoints()
        END IF
END IF
```

Revised Software for the Main Computer: The software is as described earlier except that if the current is consistently greater than a set level, $|\nabla_{Level}|$, then a collision is detected. A collision counter is reset to zero following a reading less than $|\nabla_{Level}|$ and is incremented as readings exceed the level.

If, $i_n > |\nabla_{Level}|$ then; $count_n = count_n + 1$.

If, $count_n = R_{totalcount}$, then; remove Power from the joints.

Where,

i_n = Reading of the motor current for sample n.

$count_n$ = Collision counter.

$|\nabla_{Level}|$ = Set level.

$R_{totalcount}$ = Set number of readings before detection was assumed.

The level is monitored and, once exceeded, an interrupt signal is transmitted to the supervisory level of the controller via the serial interface, and the machinery is stopped. An example of suitable code is as shown below.

```
IF Current >AmpsLevel% THEN
        CountLevel% =CountLevel% +1
        IF CountLevel% =StopNo% THEN
                ZeroJoints()
        END IF
END IF
```

To demonstrate the method, the path shown in figure 5.10 was input to the main computer and passed to the controller. The time taken to complete the path was initially 7.2 seconds. This was reduced to 6.2 seconds, a saving of 15%. The method worked efficiently for this example path and for all other motions without obstacles and with obvious, unnecessary and irrelevant via-points.

θ_1	θ_2	θ_3
0	0	180
50	45	135
100	90	90
150	45	180
100	0	140
50	45	100
0	90	180

Figure 5.10: An Example Robot Path.

An attempt was made to introduce dynamic obstacles into the motion improvement algorithm. The processing had to interact with the path planning procedures, and the software became complex and slow. The work was conducted in 2-D. A working system has not been achieved which could include dynamic obstacles. It is unlikely that such a system can work in real time without prohibitive processing power.

In order to test the collision detection work, collisions were simulated. A robot base was forced to stop during a move. This induced large torques in the motors and the force detection methods worked satisfactorily. Collisions were detected and differentiated from changes in direction. When the level detection algorithm was used with the method to compare velocity with demand signal, the level detection algorithm tended to detect a collision first, but occasionally collisions were detected when none occurred, due to noise in the motor current waveform.

By processing information from the currents to a motor, robot trajectories and paths can be adapted during a repeated series of moves in order to minimise current and torque peaks and thereby reduce the accelerations in the system.

Unnecessary changes in direction of the robot joints in attempting to closely follow programmed paths produced by CADCAM or teaching pendant reduce the operating speed and efficiency of machines and may excite resonances in manipulators. Using information from the currents to the dc motors, these unnecessary changes in direction can be removed.

If the joint actuator is an electro-mechanical unit and wear is important, minimising the current and torque transients reduces the mechanical forces and stresses in the equipment.

The motion improvement algorithms rely on information from the hardware to adapt a path. When paths are complex and inefficient, improvement is realised. When a path is planned by the automatic path planning systems described in Chapter seven, little improvement can be achieved.

The work described in this chapter did consider discrimination between transients due to collision and change in direction. This work has proved to be successful, but the system occasionally detected collisions when none occurred. The method could be improved by only signalling a collision when both the current level has increased above a preset limit and there is a low velocity with a large error.

The method adapted given trajectories for a Mitsubishi robot. A detailed description of the method and the initial results were presented by the author in 1987 and is included in the references. Although the adapted paths can be expected to be faster, the adapted paths do not consider obstacle constraints. The new motion is not necessarily an improvement in terms of speed or distance travelled, but the revised motions tend not to expend as much energy; accelerations are reduced or removed from the trajectory.

The method encompasses the idea that a robot can be automatically made to complete a task in a way more suited to itself rather than in a way which appears suitable to a human operator. This idea is expanded and discussed further in Chapter eight.

Figure **5.11**: Obtaining the Force Sensing Results

Chapter Six

IMAGE DATA PROCESSING AND A VISION SYSTEM

6.1 Introduction.

This Chapter discusses methods of retrieving information about the dynamic obstacles which were discussed in Chapter two and Chapter four. Chapter three described the evolution of systems which excluded a sensor sub-system and this Chapter describes the expansion of these systems to include a vision system.

The function of understanding a scene involves a complex sequence of computations. Processing must take place to raise the quality of the raw data to levels necessary to perform this analysis. A camera returns a voltage proportional to the light level of a range of points in a scene. This may be visually represented by 256 shades of grey, black through to white as displayed on a monochrome television set. These levels may not be a true representation as they are a consequence of many factors:

- The reflectance function.
- The illumination.
- The orientation of the surface.
- Mutual illumination and shadowing.
- Electrical noise.
- Visual noise.

As discussed in Chapter four, the planning system is initialised with a description of the static environment modelled as polyhedra. Obstacles were detected by the vision system. Several camera configurations and methods will be considered for the generation of a 3-D model: the use of a stereo image from two cameras, using a view with a single camera set at an angle to the work-place, using an overhead camera with pattern recognition to detect the obstacle from a set of known obstacles and using an overhead camera to establish the area in the X, Y plane and then setting the height to infinity for all obstacles.

These are discussed in Section 6.3. Low level image processing techniques are described in Section 6.4 and higher level pattern recognition techniques are described in Section 6.5.

6.2 Overview of an Apparatus (Including a Vision System).

The system hardware described in Chapter three needed a Vision Computer and would be as shown in figure 3.1.

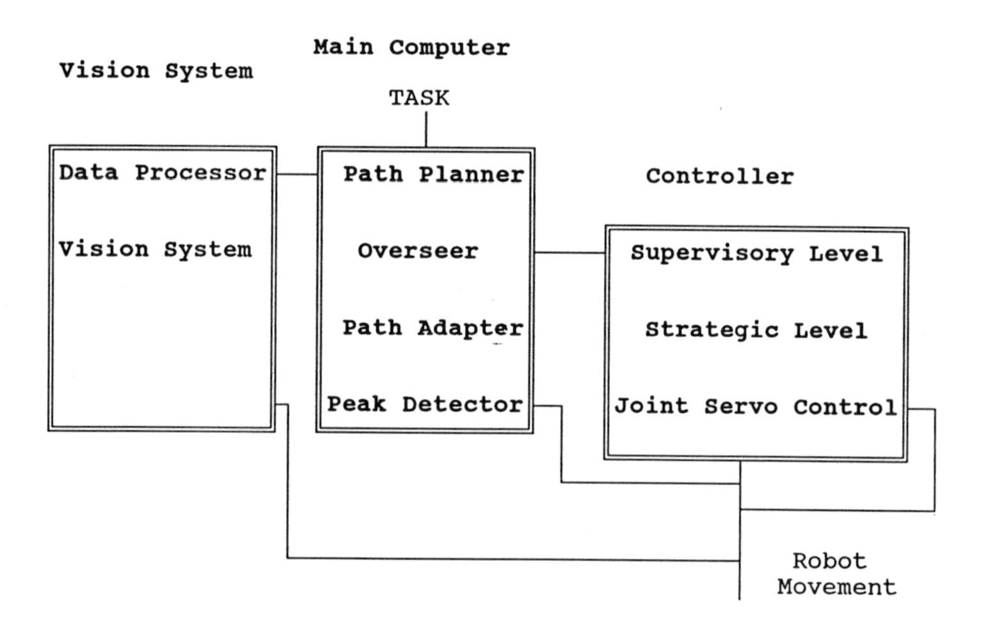

Figure 6.1: A Simplified Block Diagram of a complete system.

The data processing module responsible for the transformation of obstacles into joint configuration space would be better placed in the vision system computer. This would allow the three main sub-systems described in Chapter three to operate in parallel. That is:

- The detection and modelling of the Obstacles.
- The Path Planner and Path Adapter.
- The Robot Controller.

At least a dedicated 80286 micro-computer with a co-processor is required for the vision system. A plug in interface card is needed for each camera; typically Hitachi standard 625 cameras. The Digihurst MicroEye interface card is a simple but slow A/D converter for the analogue camera data. In its delivered state this board can take 20 seconds to capture an image of 256 x 160 pixels. After simple modifications to the hardware this time can be reduced to between 4 and 5 seconds.

Each interlaced frame from a camera takes 20 ms. With a 625 line screen and an aspect ratio of 4:3, each line takes 64μs and each picture element 0.14μs. This requires an A/D converter to transform within 0.14μs (7.25 MHz); the A/D supplied with the converter board (ZN427) operates with a minimum conversion time of 10μs (0.1 MHz). This is below the rate necessary to operate a real-time digital display and this simple system is restricted to retrieving one line scan of data during each frame. As the work described in this book concentrates on Automatic Motion Planning and not data acquisition, this data rate may be considered adequate. In order to run the system in real time a more suitable but more expensive option is an Electric Studio frame grabber.

In both cases the interfacing with the camera A/D boards is best written in assembly language.

Lighting: One or more lamps positioned over and around the work cell can provide sufficient light. A diffusing element composed of finely woven fibre can be placed over the bulbs to evenly distribute the illumination. Any real-time system will suffer from varying light conditions and noise. Attempts can be made to eliminate this problem by flooding the area with illumination bright enough to submerge shadows and noise. Even in these conditions the camera may still require repeated calibration.

A calibration technique can be used employing an initial scan which is free of obstacles. Throughout this scan, a tally can be maintained on the highest and lowest gray levels in the scene. These ideally should be 255 and 0, but due to lighting conditions may typically be between 110 and 0, less than 50% of the available range. The maximum and minimum values having been found from the initial scan, the contrast can be stretched over the full dynamic range of 0 to 255. That is,

$$I_{NEW} = \frac{(I_{OLD} - I_{min}) \times 255}{(I_{max} - I_{min})}$$

where $\mathbf{I}$ =Pixel intensity.

The reference frame having been stored, the scene can be continuously re-scanned and each element subtracted from the reference frame. This has the effect of removing the effects of uneven illumination caused by reflections external to the scene. In practice noise points may subtract to give negative results and small differences may give misleading results. To obtain a definite image a back lighting system may be required.

6.3 Obstacle Detection: The Configuration of the Apparatus.

The four methods introduced in Section 6.1 are discussed in this Section.

(a) A stereo image from two cameras.

(b) A view with a single camera set at an angle to the work-place.

(c) An overhead camera with pattern recognition software to detect the obstacle from a set of known obstacles.

(d) An overhead camera but setting the height of obstacles to infinity.

(a) Stereo imaging: Two cameras positioned above the work place as shown in figure 6.2.

Two images can be captured and stored in separate arrays. After edge detection as shown in figure 6.3, template matching methods are used to attempt to recognize the common features in both images.

If the work is limited to black obstacles against a white background or silhouettes, this problem can be made more simple than the general template matching problem.

This is similar to considering obstacles introduced into a known background where different images are compared and the methods are similar to those presented by Lo in 1990 and referenced in this book.

Once obstacles have been recognised in both cameras the position and height of the object can be calculated. The distance from the cameras to the obstacle is determined by comparing the corner points of the obstacle in the two images. The obstacle height can be calculated by subtracting this distance from the height of the cameras.

The two cameras must be a known distance apart. The corner points of the obstacle appear on both images. If these images overlap, the points do not coincide. The position in the image and the distance between the points can be used to determine the range of the obstacle from the cameras and therefore the height of the obstacle. As obstacles are moved closer to the cameras the disparity will become less. The system is calibrated to determine the height of the obstacles.

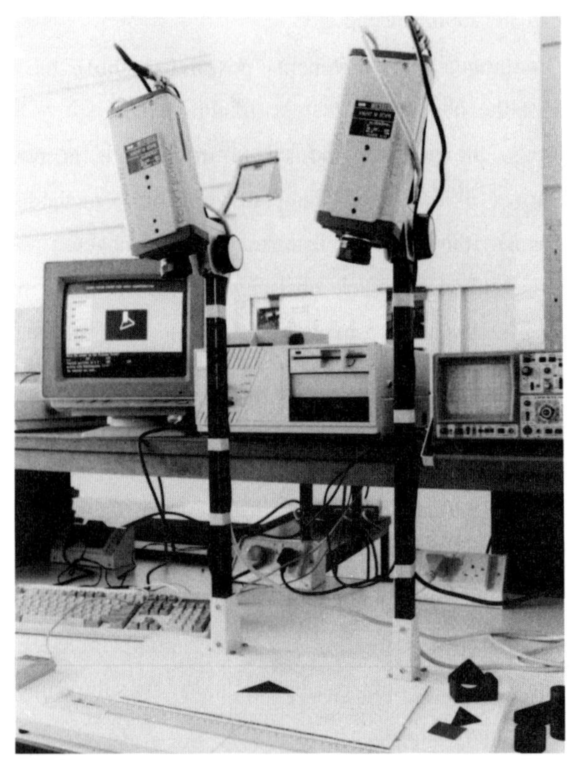

Figure 6.2: The Configuration for the Stereo Vision System.

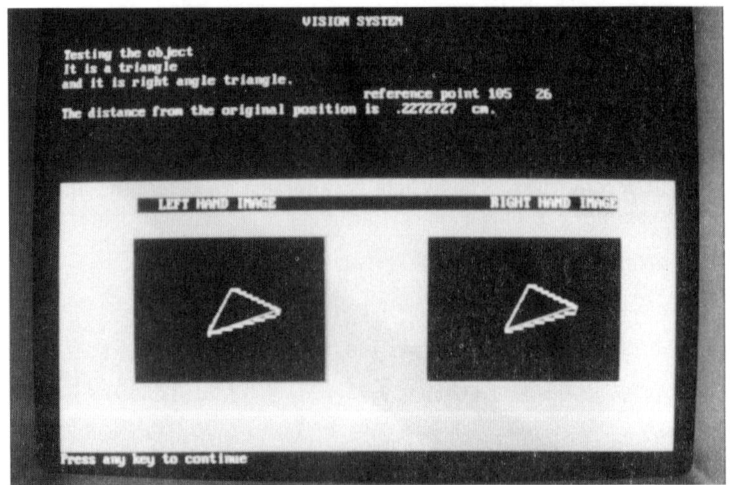

Figure 6.3: The Stereo Image after Edge Detection.

The range can be approximated by:

$$range = \mathbf{r} = d\{\sqrt{(f^2+x_L^2+x_R^2)} \div (x_L - x_R)\}$$

and the height of the object can then be found: *height* = **h** = h_c - **r**

where **r** = Range from the left camera lens if the point was in the right side of the scene.

or

= Range from the right camera lens if the point was in the left side of the scene.

d = Distance between the camera lens centres.

f = Focal length of the cameras.

x_L = Distance of the image pixel from the left centre position.

x_R = Distance of the image pixel from the right centre position.

h = The height of the object.

The values of x_L and x_R can be positive or negative depending on their location relative to the centre of their respective images.

(b) Single camera viewing from an oblique angle: This is shown in figure 6.4. A strong source of illumination is placed directly behind the camera and a diffused source placed at 90°. This configuration enables faces of an object to be illuminated with different levels of incident light. This produces distinct gray regions separating the faces. The calibration involves a non-linear relationship as the size of the object will vary with distance from the camera and some form of complex pattern recognition is required.

A program was written by the author to produce a 3-D model of obstacles by fitting lines (y=mx+c)to data produced by a trace routine. Each fitted line was stored in an array for later comparison as a gradient(m) and a Y crossing point (c). Code was also inserted to allow lines of infinite or zero gradient to be fitted. The method is discussed further in the appendices and is based on work described by Oaten in 1990. Due to the inaccuracies inherent in fitting data points and the fitting percentage of the line, several lines of similar gradients and crossing points were generated. A double sort routine was used to select similarly proportioned lines which were averaged to produce single lines corresponding to the straight edges of the object. Figure **6.5** illustrates the results of the line fitting procedure on a 3-D object (triangular prism).

Figure 6.4: **Single Camera viewing from an Oblique Angle.**

From this mathematical representation, it is possible to generate a polyhedral model of an object. That is, the absolute positions of the corner points can be labelled and stored as a three-dimensional wire frame representation including vertex labelling to indicate each edges' contribution to the complete object description. Three lines crossing at a point represent a visible corner, two crossing lines at a point represent corners where a third (or greater number) edge is hidden from the camera. Visible corners vertically above one another can be assumed to approximate to height. Although 3-D spatial information can be gleaned from this 2-D picture, edges hidden from view are unknown quantities. The programs developed demonstrated the large amount of computation time and long code length that was necessary to analyze a relatively low resolution 3-D image.

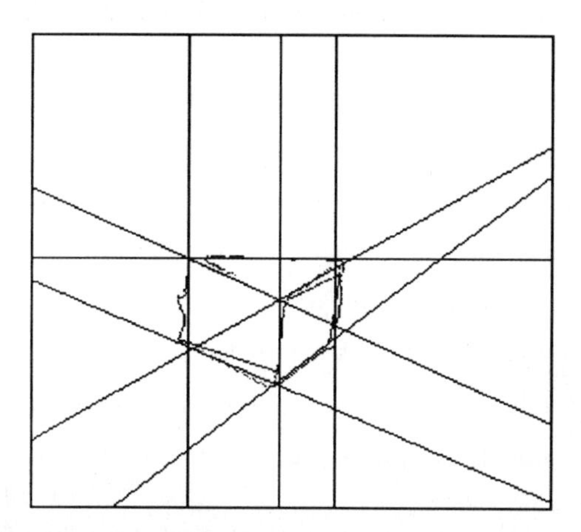

Figure 6.5: Line Fitting for a 3-D Triangular Object.

(c) Single camera above the work place: (Using Pattern Recognition): Since the camera is fixed, the X and Y coordinates are calibrated to refer the physical position to the array position. 1.55 mm per vertical pixel and 1.85 mm per horizontal pixel can accurately position an object under the camera. A typical configuration is shown in figure 3.1.

(d) Single camera above the work place: (Infinite Height): As described above, the camera being fixed enables calibration of the X and Y coordinates to refer a physical position to the array position. In this case the height of the obstacles need not be determined as the models of the obstacles are given an infinite height. This method can be used if template matching routines do not work in real time, or fail, or the paths must be safe and "outside" the obstacle.

Discussion: Each of the methods above is considered and discussed individually:

(a) Using a stereo image from two cameras. Once the stereo system has been set in place, the accuracy is dependent on the relative positions of the matched image points (the Correspondence problem). The method attempts to match points using edge detection techniques and by searching for corner points. This can be achieved by creating a window around corners in one image and searching for similar areas in the other image. The method is complex and often unreliable. For the prototype system developed at Portsmouth Polytechnic, the processing took up to 90 seconds while still not guaranteeing a result. The method requires much programming time and faster or parallel processors to produce a usable system, and this is discussed later in this book.

(b) Using a view with a single camera set at an angle to the work-place.
This method is very susceptible to changes in illumination of the scene. This requires careful adjustment. Lighting aberrations and shadowing cause malfunctions in the software routines and often this results in shadows being mistaken for regions of interest and being analyzed as objects.

When strong light is provided from directly behind the camera, the shadowing effects are minimised or moved to the rear of the object being viewed and thus partially hidden from the camera. This will reduce but often will not remove the possibility of shadow regions being analyzed as objects.

These systems will often malfunction and the processing is more complex than for any other configuration of the apparatus.

(c) Using an overhead camera with pattern recognition techniques.
This configuration of the apparatus tends to give the best results and allows

the simplest processing techniques to be used. This method is assumed in the description of the rest of the systems in this book. The pattern recognition techniques are discussed in Section 5.5.

(d) Using an overhead camera, setting the model height to infinity.
This method will dramatically reduce the available free space (as shown in figure 4.40) but is useful when pattern recognition techniques fail to identify an object from a set of templates.

Conclusions: The configuration described in (c) is useful. The use of templates in real time requires the vertically mounted camera without the distortion problems associated with the obliquely mounted camera. In the latter case the size of the object varies with its distance from the camera. In both (a) and (b) the processing time will be excessive.

Use of technique (d), setting the obstacle to infinite height, allows an alternative system to continue operating in the presence of unexpected obstacles or if errors occur.

6.4 Obstacle Detection: LowLevel Vision Techniques.

Processing at a low level reduces the effects of noise and shadowing prior to reducing the size of the arrays and applying the techniques of edge detection and pattern recognition. There are standard techniques to enhance a raw image in order to enable accurate sizing and recognition of an object under a camera.

The methods described are based on spatial domain techniques; that is, methods that operate directly on the pixels in an image. These can be expressed as:

$$g(x,y) = \mathbf{T}\,[f(x,y)]$$

where

$f(x,y)$ is the original image.
$g(x,y)$ is the processed image.
$\mathbf{T}$ is an operator over some neighbourhood of (x,y).

The following methods are considered and are discussed in this section:-

(a) Gray level weighting.

(b) Smoothing.
  (i) Neighbourhood averaging.
  (ii) Weighted neighbourhood averaging.
  (iii) Median filtering.

(c) Thresholding.

(d) Reduction of the array size.

(a) Gray Level Weighting: The data obtained from a camera provides explicit information regarding the gray level content of a scene. For object recognition it is more useful to enhance information regarding an object and to reduce the gray levels referring to the background. The implementation of such a process involves an overall loss of information, although the loss of irrelevant data is offset by an increase in the relevant foreground gray levels.

By weighting the data, closely matched gray levels can be separated and this helps to organise the data in preparation for thresholding procedures. The method works but is time consuming. The process operates by calculating a histogram of the gray levels in a sampled image. Specifically it calculates the frequency of occurrence of each of the separate gray levels. A sample histogram of a work place with an obstacle present is shown in figure 6.6.

The histogram shows a high incidence of "white" (background) and a low incidence of dark gray (object). By flattening this histogram to a level at which the incidence of white was lowered and the incidence of gray was raised, an enhanced image is produced. The histogram flattening is performed by dividing the sum of the histogram values up to each gray level into a large number of increments for relatively frequent values and smaller values for rarer gray levels.

A program to perform gray weighting totals the quantity of gray levels up to the maximum (usually $\approx$110) and typically divides this into sixteen equal cumulative frequency distributions. These individually aggregated frequencies having been set, the histogram is consulted and a new addition of frequencies begun.

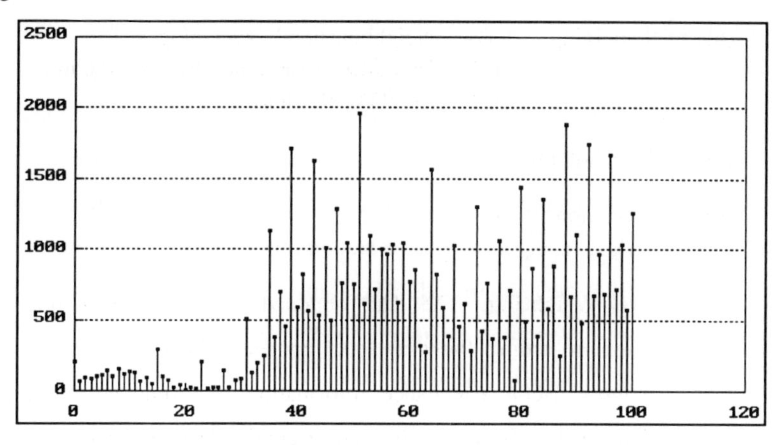

Figure 6.6: A Sample Histogram of the Raw Image Data.

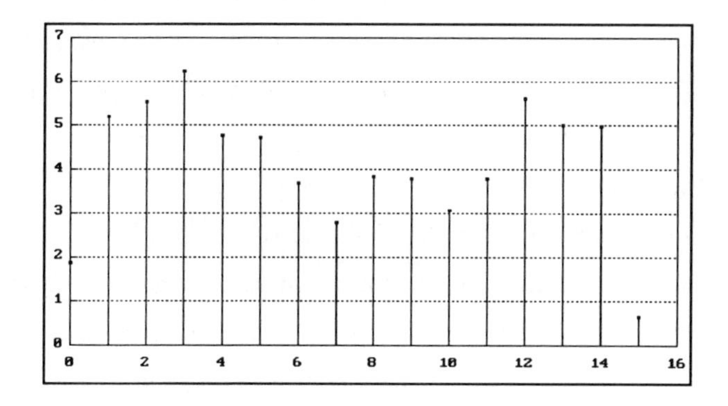

Figure 6.7: A Histogram of the Adjusted Image Data.

When the quantities of frequencies equals the individual frequency total, the gray level reached is marked and the process repeated from the marked gray level. The levels are more widely spaced in the regions of rarer frequencies. A histogram of this new adjusted sample image is shown in figure 6.7.

(b) Smoothing: Smoothing is a technique applied to raw pixel data in order to remove noise. Smoothing algorithms tend to blur an image, especially at edges where there are abrupt changes in intensity. This can be related to the frequency plane, where edges imply high frequency components which are smoothed using the low pass filter that many algorithms emulate. All smoothing algorithms in the spatial domain compare the value of a pixel with its neighbouring pixels and, using some form of interpolation, replace the pixel in question with a smoothed value. This can be particularly effective when applied to individual 'spot' noise. Three smoothing algorithms are considered:

(i) Neighbourhood Averaging.

(ii) Weighted neighbourhood averaging.

(iii) Median Filtering.

(i) **Neighbourhood Averaging:** Given an N x N image f(x,y),the procedure generates a smoothed image g(x,y) whose gray level at each point in an image (x,y) is obtained by averaging the gray level values of the pixels contained in a predefined neighbourhood of (x,y), say (m,n). That is:

$$g(x,y) = 1/(m \times n)\ \Sigma f(m,n)$$

That is,for a 3x3 neighbourhood, the centre pixel in the window shown below is replaced by the average of the pixels in the 3x3 window.

(x-1,y-1)	(x,y-1)	(x+1,y-1)
(x-1,y)	(x,y)	(x+1,y)
(x-1,y+1)	(x,y+1)	(x+1,y+1)

Figure **6.8**: Neighbourhood Averaging.

The centre pixel's new value becomes:

[value(x-1,y+1)+value(x,y+1)+value(x+1,y-1)+value(x-1,y) +value(x,y) +value(x+1,y)+value(x-1,y+1)+value(x,y+1)+value(x+1,y+1)]/9

(ii) **Weighted Neighbourhood Averaging:** A typical technique is to weight only the centre pixel of the window and then to average as above. In practice the best results are often to be achieved with a weighting of approximately eight.

(x-1,y-1)	(x,y-1)	(x+1,y-1)
(x-1,y)	8*(x,y)	(x+1,y)
(x-1,y+1)	(x,y+1)	(x+1,y+1)

Figure **6.9**: Weighted Neighbourhood Averaging.

The centre pixel's new value is:

[value(x-1,y+1)+value(x,y+1)+value(x+1,y-1)
+value(x-1,y) +value(8.(x,y) +value(x+1,y)
+value(x-1,y+1) +value(x,y+1)+value(x+1,y+1]) /16

(iii) **Median Filtering:** The same 3x3 window of pixels is used but the median of the values is selected as the new value for the centre pixel. As an example, the centre pixel and its 8 neighbours are shown in figure 6.10 below. 0 refers to black, whilst 15 refers to white.

15	2	2
3	9	10
10	10	12

Figure **6.10**: Median Filtering.

The values are numerically ordered and the centre (median) value selected. The median value from figure 6.10 is underlined:

2 3 3 9 <u>10</u> 10 10 12 15

Selection of the smoothing method: Using the averaging method many noisy "spot" pixels may be wrongly converted into small regions and the edges of obstacles tend to become excessively blurred. This causes problems when accurate sizing of the object is required for object recognition.

Median filtering is computationally slower but produces better results in terms of less blurring and an almost total elimination of "spot" noise. Any spot noise is moved to the high end of the list, and is not selected. A standard "bubble sort" procedure can be used to select the median value. Although a "quick sort" procedure is usually faster for large unsorted arrays, for only nine elements the bubble sort tends to perform the operation more quickly.

(c) Thresholding: Later problems of analyzing abrupt changes in gray intensities could be compounded by the smoothing operation which tends to soften sharp edges into ramp functions. Thresholding is used to segment an image into regions of similar gray levels. A threshold level is set and pixels are compared with this level. Pixels above a threshold are set to one value, for example 1, and pixels below the threshold are set to a different value, for example 0.

That is:

$$\text{If } f(x,y) \leq T \text{ then } f(x,y) = 0$$
$$\text{If } f(x,y) > T \text{ then } f(x,y) = 1$$

where T =Threshold Level.

(d) Reduction of the Array Size: Processing time is a function of the size of the image array being processed. The image array can be reduced by only considering the area containing an object. The technique is to quickly scan the image data until an object is detected. The routine then focuses on the object to obtain the maximum amount of information concerning the area of interest.

The method checks the number of pixels below the threshold level after each column scan. When the number of pixels is greater than two and an obstacle has not already been detected, then the column scan steps are reduced to give a higher resolution and a flag "OBJECTDETECTED" is set to true. The row of the image array is set and an "ALLCLEAR" flag which may be used to signal the controller is set to false. This may be achieved using the following code

```
IF NoPixels >=2 AND ObjectDetected% =false% THEN
      ColumnSteps% =2
      OffsetRow% =Row%
      ObjectDetected% =true%
      ObjectRow% =ObjectRow% +1
      AllClear% =false%
      CALL ColumnScan          ; load image data into array
END
```

While the image is being scanned the pixels are tested against the threshold level and once the scan has left the obstacle the column scan steps are increased. The "OBJECTDETECTED" flag is set to false so that data is no longer loaded into the new smaller array.

```
IF NoPixels% <=4 AND Object Detected% =true% THEN
 ColumnStep% =4
 ObjectDetected% =false%
END
```

The array size becomes the object row size x the number of columns. The limits of the columns which contain the object can then be found in a similar manner.

6.5 Obstacle Detection: High Level Vision Techniques.

Several parameters may be considered for the matching of templates used for recognition, but before these are applied, edge detection techniques are used to locate the boundaries of separate regions. The process detects abrupt discontinuities within the image and uses a local derivative operator to transform these discontinuities into marked edges. The gradient operator used to detect edges is defined as the two-dimensional vector $G\nabla$ such that:

$$G\nabla = [G_x, G_y]^T$$

The gradient is defined as $|G_x| + |G_y|$ where

$$G_x = \delta f / \delta x$$

$$G_y = \delta f / \delta y$$

In digital form this is the difference in intensity of horizontal and vertical neighbours of the pixel under scrutiny using the first order difference:

$$G_x = f_x(i,j) = f(i+1,j) - f(i,j)$$

$$G_y = f_y(i,j) = f(i,j+1) - f(i,j)$$

Once the edges have been detected templates are produced to match the images against relevant details of obstacles held in computer memory. Several types of template matching algorithms are described in the literature including Groover et al in 1986, Fu et al in 1987, Fairhurst in 1988 and most recently Galbiati in 1990. Three are considered here:

(a) Fixed size parameters - area, perimeter etc.

(b) Rubber band parameters - internal angles, length of sides ratio etc.

(c) Mathematical transformations (eg the HOUGH transform).

Fixed templates are used as a set of known obstacles is assumed and typical parameters selected for matching procedures are:

(i) **Area** *Obtained by counting the pixels with a common property.*

(ii) **Perimeter** *Obtained by counting the connected edge pixels.*

(iii) **Diameter** *The maximum distance between edge points around an object.*

(iv) **Compactness** **=(Perimeter)^2/Area**

(v) **Thinness** **=Diameter/Area**

These are described:

(i) **Area:** During the thresholding process an Area count can be inserted and incremented whenever a pixel is identified as part of an obstacle. That is:

```
IF f(X, Y) <T THEN
  Area =Area +1
END
```

(ii) **Perimeter:** A similar process can take place during edge detection. Whenever a pixel is identified as being part of an edge the perimeter count is incremented. That is:

```
IF (Gradient X +Gradient Y) >GradientLevel THEN
   Perimeter =Perimeter +1
 END
```

(iii) **Diameter:** During edge detection the positions of the edge pixels can be stored in an array. On completion of the edge detection process each of these positions may be compared to the other positions stored in the array. The largest distance between any two edge points is the diameter. To calculate the distance between edge points the row value of the pixel under test is subtracted from the row value of the reference pixel and named, for example "endx". A similar process is completed for the column values to produce a reference pixel "endy". The distance between the edge points is calculated and the largest distance is recorded as the diameter. This is the maximum distance between any two pixels in the obstacle shape. Typical algorithms are:

(If RefX and RefY identify the recorded edge pixel and X and Y identify the edge pixel under test), then

```
endx =(X - RefX)
endy =(Y - RefY)
TempDia =√(endx² +endy²)
IF TempDia >diameter THEN
    diameter =TempDia
END
```

(iv) & (v) **Compactness and Thinness:** These are ratios derived from the Area, Perimeter and Diameter such that:

$$\textbf{Compactness} = \textbf{Perimeter}^2 \,/\, \textbf{Area}$$

and

$$\textbf{Thinness} = \textbf{Diameter} \,/\, \textbf{Area}$$

Once the parameters are found, two alternatives are considered for pattern recognition:

(a) Probability.

(b) Average error.

(a) **Probability:** To calculate the probability of an object being one of a set of known objects the parameters are taken as a percentage of error with respect to the template parameters. The error for each parameter for each template is calculated and summed and the error for each template parameter is divided by the summed error of that parameter for every template. This gives a probability value of error for that template for each parameter. When all the probability errors are added together a value of one is the result. The probability of an error occurring for each template is found by taking the mean probability of all the parameters for the particular template. Typical equations are shown below:

$$\text{Parameter Error} = \frac{(\text{Object Parameter} - \text{Template Parameter})}{\text{Template Parameter}}$$

$$\text{Probability Error} = \frac{\text{Parameter Error}}{\Sigma \text{ Parameter Errors}}$$

$$\text{Average Prob of Template} = \frac{\Sigma \text{ Probability Errors}}{\text{Number of Parameters}}$$

For interpretation, it is assumed that if the computer could not recognise an object then the probability of an error for any template is 1/Number of Templates. This value may prove to be too high for interpretation and may need to be reduced to half the value, that is: Interpretation value = 1/(Nx2)

Although the error for all the templates could be large for an unrecognisable object, one error value could be less than the others and the probability of an error occurring would be less for that particular template. The probability error could be low enough to be recognised as a known obstacle.

(b) **Average errors:** The error for each parameter is found by comparing the object and template parameters. The error is taken as a percentage with respect to the template parameter. The mean percentage of the parameter errors is then found to give the error for the template and from this the lowest error is selected to find the template which matched the object.

$$\text{Parameter error} = \frac{(\text{Object parameter} - \text{Template parameter})}{\text{Template parameter}}$$

$$\text{Average error} = \frac{\Sigma \text{ Parameter errors}}{\text{Number of parameters}}$$

If the lowest percentage error of a template is less than twenty percent then it is assumed that none of the templates match and the object is unknown. This method tends to have a high success rate in interpretation and recognition once the templates have been established.

Position of the obstacle: Once the object has been recognised, the centre of the obstacle in the x, y plane needs to be calculated. The slope of the diameter is calculated and the X and Y coordinates are found for the point half way along the diameter, giving a centre position.

The X and Y positions are added to (or subtracted from) the first edge position at the end of the diameter to give an approximate position of the centre of the obstacle. That is, if endx and endy were the difference in rows and columns between two edge points and:

Xpos	=X centre coordinate
Ypos	=Y centre coordinate
RefX and RefY	=the reference edge point for the diameter
X, Y	=the opposite end of the diameter

then

```
ϴ =InvTan endy / endx
Xcentre =(diameter / 2) * COS ϴ
Ycentre =(diameter / 2) * SIN ϴ
 IF RefX >X THEN
  Xpos =X +Xcentre
 ELSE   Xpos =X - Xcentre
END
 IF RefY >Y THEN
  Ypos =Y +Ycentre
 ELSE   Ypos =Y - Ycentre
END
```

Once the obstacle has been identified and positioned, the stored model is extracted. A variable ShapeNo is initialised to the number of the matched template and an array with the parameters of known and expected obstacles is consulted to provide the Z coordinate/coordinates. The X and Y coordinates are known from the centre of the obstacle. If the obstacle is not recognised then the parameter 'Diameter' is converted into a radius and a 'dummy' obstacle can be inserted with a large Z axis value. This prevents machinery and robots from moving over an obstacle of unknown height. Instead a robot could fold at the elbow joint and move around an obstacle.

Detection of Movement: The row and column number in which the object is first detected (RowOffset% and StartCol%) is noted once the image has been captured. Limits are set around the values to allow for changes due to lighting, shadows or noise. During subsequent scans RowStart and ColStart can be tested against the limits set by the last scan. If they are outside the limits then it may be assumed that the obstacle has moved and reprocessing takes place to find the new position. This is shown below:

```
****** Testing the Obstacle Position against the Limits ******
    IF StartCol% >Mov(ColPos%) OR StartCol% <Mov(ColNeg%) THEN
        Moved% =true%
    IF RowOffset% >Mov(RowPos%) OR RowOffset% <Mov(RowNeg%)
        Moved% =true%
****** Loading in new Limits before object reprocessed ******
    Mov(Cpos%) =Limits%(Cend%) +1
    Mov(Cneg%) =Limits%(Cend%) - 1
```

The Detection of Multiple Obstacles: If two obstacles are detected, in order to store their image data separately, a variable RowImageNo is initialised and incremented each time an object is detected. The row and column limits of each image can be stored in a small array.

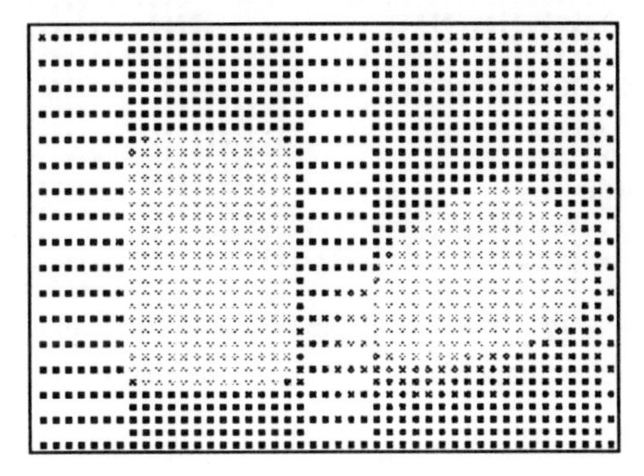

Figure 6.11: Polyhedron and cylinder (detected by a vision system).

Obstacles could also appear above and below one another. An obstacle above and to the left of a lower object would cause the end column limit of the upper image to increase as the program took the second image to be part of the first. The end column limit can be stored in a temporary array and tested with the last end column scan limit. If the difference is greater than a preset limit then it is assumed that the first image has ended. The information from the second image can be recorded seperately as shown in the following code:

```
IF Temp%(ColEnd%, Colimages%) <column% THEN
 Temp%(ColEnd%, Colimages%) =column       ; Up dating end column limit
Difference =(Temp%(ColEnd%, 1) - Limits%(ColEnd%, 1))
  IF Difference >5 THEN
  Transfer limits between array locations
 Limits%(ColEnd%, Colimages) =Temp%(ColEnd%, Colimages%)
```

A similar routine can deal with obstacles above and to the right of a lower obstacle as shown in figure 6.12. In this case the start column limits are tested. If the difference of the present limit compared to the limit of the last column scan is less than eight it is assumed that a second object has been detected. As before, the limits obtained from the first image are then transferred to the second image location.

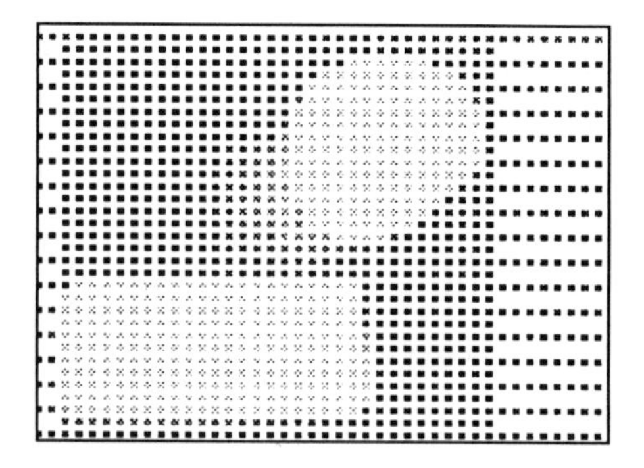

Figure **6.12**: The Obstacles in Alternative Positions.

The image processing and pattern recognition routines are contained within a FOR, NEXT loop so that both obstacles are pattern matched and modelled.

6.6 Results.

A Screen Display: In the example shown in figure 6.13, a camera was set 0.6m above the floor of a work cell (the top of a work bench) and the object was a polyhedron (an oblong box).

Figure 6.13 shows the identification of the object, the size of the object, the diameter and positional information. The changes in the image array after filtering, thresholding and edge detection can be seen with the matched template and percentage of error.

Recognition timings: The typical times taken for recognition of a horizontal cylinder using different sizes of array were recorded and are shown below. The processing included smoothing, thresholding, edge detection and template matching.

The results in figure 6.15 were obtained using array reduction and are a comparison of the different timings for the different obstacles considered. The obstacles were placed with their longest length across the image array.

The processing times and the number of BLOCKED nodes detected were recorded for various obstacles in a variety of positions. The positions where the data were recorded are shown in figure 6.16 and are reproduced from figure 4.6.

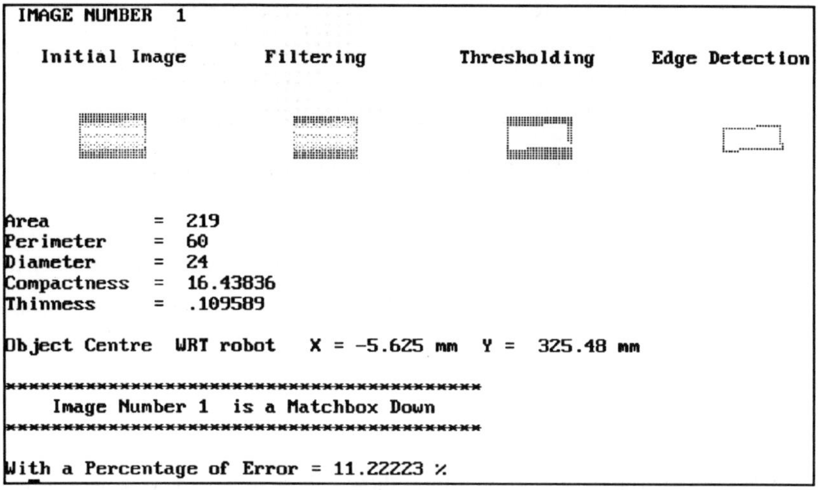

```
IMAGE NUMBER  1

  Initial Image        Filtering         Thresholding      Edge Detection

Area          =  219
Perimeter     =  60
Diameter      =  24
Compactness   =  16.43836
Thinness      =  .109589

Object Centre  WRT robot   X = -5.625 mm   Y =  325.48 mm

******************************************
    Image Number 1   is a Matchbox Down
******************************************

With a Percentage of Error = 11.22223 %
```

Figure 6.13: Identification of the Polyhedron.

Array Size	Rows x Columns	Processing Time
440	22 20	1.4 seconds
2080	52 40	1.9 seconds
13,312	104 128	7.5 seconds

Figure 6.14: Table of Recognition Timings for different Array Sizes.

Processing times: The typical timings shown below in figure 6.17 included the recognition of the obstacle, transformation into discrete 3-D joint space and the transmission of the BLOCKED nodes to the path planning computer. Times are in milli-seconds.

Obstacle	Processing Time (milli-seconds)
Horizontal Cylinder	950
Vertical Cylinder	450
Horizontal Polyhedra (Maximum Area)	920
Horizontal Polyhedra (Minimum Area)	910
Vertical Polyhedra	470
Cube	750

Figure 6.15: Table showing typical recognition times for various obstacles.

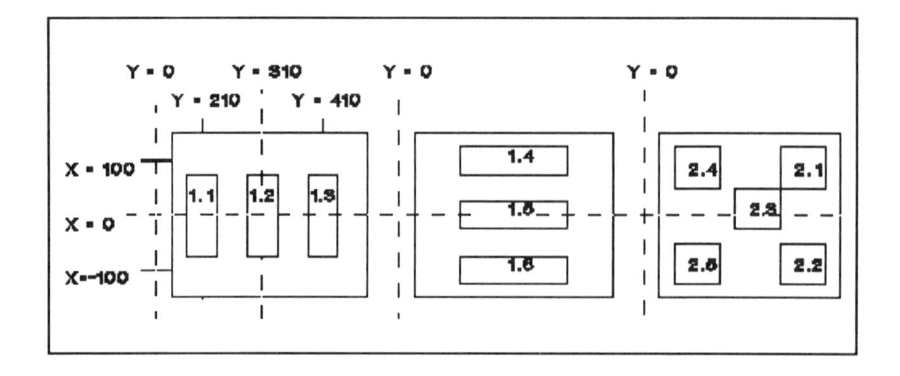

Figure 6.16: Reference Positions.

Reference Position	**Horizontal Cylinder**	**Horizontal Polyhedra** (Minimum Area)	**Horizontal Polyhedra** (Maximum Area)
1.1	4.85	3.41	4.18
1.2	4.8	3.64	4.15
1.3	4.65	3.59	3.81
1.4	4.85	3.69	4.8
1.5	4.91	3.79	4.59
1.6	4.78	3.58	4.26

Figure **6.17**(a): Processing Times for the Obstacles with larger areas in the X,Y plane.

Reference Position	**Vertical Polyhedra**	**Vertical Cylinder**	**Cube**
2.1	3.48	4.56	3.78
2.2	3.78	4.78	3.48
2.3	4.11	4.98	4.11
2.4	4.63	5.6	4.63
2.5	4.52	5.7	4.52

Figure **6.17**(b): Processing Times for the Obstacles with smaller areas in the X,Y plane.

6.7 Discussion and Conclusions.

The programs discussed use various processing techniques to provide simple pattern recognition through the use of template matching. Data in a form capable of performing 3-D manipulation are then generated from the 2-D images. Problems may be overcome by using mathematical image enhancement techniques applied to the image information.

Amongst the mathematical techniques to enhance the image, frequency analysis is an important concept, but it is difficult to use in a real-time system because of processing limitations.

Methods of recognition and the transformation of objects into a robot joint space have been discussed. Data in a form capable of being analyzed by a motion planner and path improver have been described. Programs can use a vertically mounted camera to analyze objects in the work space as binary images. The data are pre-processed and then used for template matching. If sphere models are used then the object may initially be modelled by a minimum bounding sphere, if time allows, a more accurate spherical representation can be used.

Parameters (iv) and (v), (compactness and thinness), may be used to distinguish between obstacles of similar size but of a toroidal nature. They are also useful when shadow effects alter the absolute size of an obstacle as the ratios tend to be more repeatable than for the other parameters.

An obstacle can be thought of as "identified" when it satisfies the minimum error criterion set within the software. It is possible for an unknown object to satisfy the conditions set, and to be incorrectly identified. The programs should always correctly identify the maximum diameter and the minimum bounding cylinder necessary for obstacle avoidance without the need for this recognition processes.

The reduction in the array size can reduce recognition times from >7 seconds to <1 second. This is demonstrated for the horizontal cylinder in figure 6.14 using a forced array size. Using reduced array sizes selected by the system, the timings for different obstacles positioned in the same place in the X,Y plane are shown in figure 6.15. The objects with a larger area tend to take a longer time, as the array sizes are larger.

Typical processing times are shown in figures 6.17(a) and 6.17(b). The timings varied from 3.41 seconds for a polyhedron with minimum area showing to 5.7 seconds for a vertical cylinder.

Large obstacles, for example the cylinder, tended to take longer to transfer into 3-D joint space.

Chapter Seven

PATH PLANNING

7.1 Introduction.

The different aspects of the work described in Chapters three, four, five and six are combined in this Chapter to describe an advanced automated work cell. Within this cell, a robot is able to move under a camera which views a large section of its work space. The image produced by the camera is captured by a computer and analyzed. The analysis produces a list of configurations which are blocked to the robot within the cell.

This Chapter describes the evolution of a motion planner. The problem involves moving a robot from one place to another while avoiding collision with obstacles. Initially the problem is considered in 2-D space for a robot base and a simulated joint and link. Later in this Chapter the work is extended to 3-D space for use with a Mitsubishi RM-501 robot.

START and GOAL configurations are entered by a human operator. The path planning computer uses these configurations and the obstacle data to calculate a path for the robot to move safely through the field of vision of the camera. A third computer uses this path to direct the movement of the robot around the work-cell. Only a section of the robot work area is covered by the vision sub-system,and other sub-systems would be required to cover the whole work area.

The position problem is solved for the START and GOAL positions using the forward kinematics solution presented in Section 3.7. The result specifies how much each joint has to be rotated to effect the desired movement. This is the initial path. When no obstacles are detected by the vision system the direct trajectory locus from START to GOAL is automatically selected.

The control flow for the robot system is thus:

Task
Set by a human programmer and entered into the main computer.
|
Trajectory locus
Calculated by the path planner in the main computer.
|
Robot co-ordinates
Extracted from the trajectory locus at the supervisory level of the robot controller.
|
Robot trajectory & robot movements
Generated by the controller.

In general, a TASK description contains information of the type:

START
Wait (for) PART
Pick PART from BELT
Put PART into BOX
Goto START

The subject of this book is concerned with improvements to motion planning methods which would allow automatic generation of the robot movements required to achieve such a TASK. In 1977 Udupa divided path planning into three stages. This type of description has been used by many researchers to describe the problem, and the path planning work described in this book uses similar stages. These are:

(i) Path feasibility.
(ii) Approach planning.
(iii) Path planning.

These stages are described:

(i) Path feasibility. The path is a series of configurations through which a machine moves in order to carry out a task. The configuration of the robot at points and the GOAL position are checked for feasibility. Positions which are outside the robot's work-space, or which would cause collisions with obstacles or the static environment, are not accepted from the human programmer.

(ii) Approach planning. Approach paths are paths which move from positions with clearance from obstacles, to GOAL positions close to surfaces. In industrial applications approach paths tend to be short. They are related to machine geometry, and are calculated for specific machine configurations. This book will not consider these paths in detail.

(iii) Path planning. The remainder of this Chapter deals with the path planning problem. To simplify the problem, path planning is initially considered for a two degree of freedom manipulator. This is described in Section 7.2. The path planning methods are then extended to 3-D space and this is described in Sections 7.3 and 7.4.

Configuration of the equipment: As described in Chapter six, the camera is placed over a section of the work place. The coordinates of the camera must be referred to the joint coordinates of the robot. An origin is defined as the centre of a base and a shoulder joint. This is used as the origin for all coordinate systems in this book.

In all cases the Cartesian coordinates are determined relative to the origin with X running from front to back of the robot base, Y running from left to right when viewed from the front of the robot, and with a vertical Z axis.

For the work described in the rest of this chapter, the camera is assumed to be positioned so that the camera base is at Y =170 mm and both Y and Z are central on the X axis.

In *Figure 7.1* the robot is displayed with the waist (θ_1) at 90°, the shoulder (θ_2) at 60° and the elbow (θ_3) at 120°.

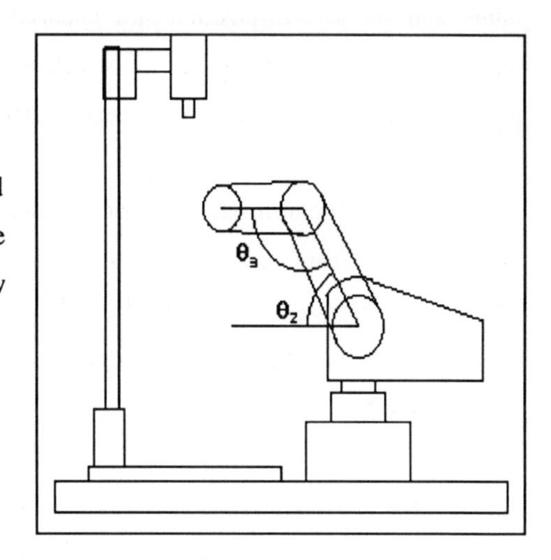

The Joint Angles.

Figure 7.2 shows a plan view of the system components when configured for use with a camera base and a front lighting system.

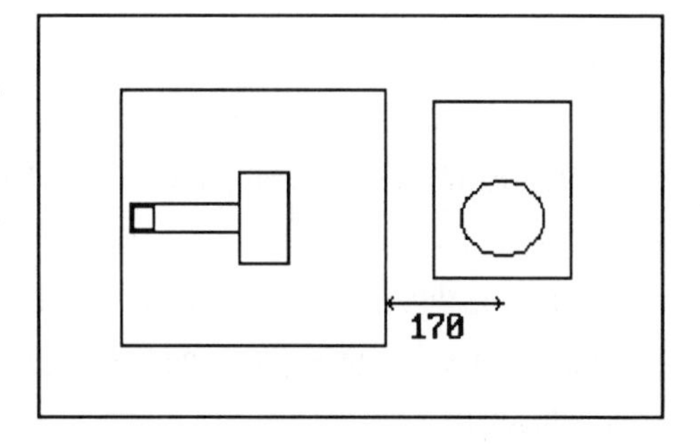

Plan View.

7.2 Path Planning for a Single Link Manipulator.

Simple generalisation from the 2-D problem to the 3-D problem is not possible, but solving the 2-D problem is a useful introduction to the general path planning problem. The two degree of freedom manipulator considered in this Section consists of two joints co-located at the origin; a base joint θ_1 and a simulated shoulder joint $\theta_{2\underline{sim}}$ and a simulated single link $L_{1\underline{sim}}$.

Two obstacle models are considered during this introduction in 2-D space; a sphere and a simple parallelepiped (a solid bounded by parallelograms).

Statement of the 2-D problem. The robot is to move through a set of via points from a START to a GOAL configuration avoiding obstacles and without violating geometric constraints. Without considering orientation, the purely position problem, $p_{(t)} = p_{(X1,Y1\ X2,Y2\ X3,Y3} \ldots$etc) can be stated for a move from one position to the next as :

From

$$P_n = (X_n, Y_n), \quad \text{that is} \quad P_n = (\theta_{1(n)}, \theta_{2(n)})$$

Move to

$$P_{n+1} = (X_{n+1}, Y_{n+1}), \quad \text{that is} \quad P_{n+1} = (\theta_{1(n+1)}, \theta_{2(n+1)})$$

Where

P_n is the nth position in space.
$\theta_{i(j)}$ is the jth position of joint i in a trajectory locus.
X and Y are Cartesian coordinates.

Two solutions to this 2-D problem are considered:

(a) A local and heuristic method.
(b) A global method.

(a) **A local and heuristic method:** If the data for the world model is stored on a floppy disc, a file containing simulated obstacles can be loaded into the vision computer, bypassing the camera input. Each TASK is input by a human operator and consists of the initial, intermediate and final coordinates of the ForeTip. The path trajectory locus is calculated from the TASK description and the model data. The trajectory locus consists of robot coordinates and these are down loaded to the robot controller.

An example of the approach path could be defined as follows:

(i) Position the GOAL 10 mm above the final position.
(ii) Move down in a straight line at 1/3rd normal speed.
(iii) Simulate gripping the part.

When moving away from the final position a similar motion is used. A part is simulated being lifted 10 mm at 1/3rd normal speed. The approach paths are defined by a few lines of program code written by the human operator. If a new approach path is required it is simple to modify this code.

The START configuration is the first node on a graph. The path cost function is set to 0 for the START node and $FFFF_H$ for the other nodes. Paths around obstacles are represented by nodes for both the sphere and parallelepiped models.

After the graph has been initialised for searching, the direct path from START to GOAL is tested. If this path is blocked then the algorithm selects new nodes until either the GOAL is reached, or all nodes have been tested. If all nodes are tested and no path is found then it is assumed that no path exists.

From the START node, paths are considered to all the other nodes. Each of these paths is tested for collisions with obstacles. If a path is clear, the cost of the path is calculated. If the cost of the path to a new node is less than any previous path then the new cost and the previous node are stored for the new node. Once each node has been tested, the node is recorded on a list so that the node is not retested. This list forms the trajectory locus to be passed to the robot controller.

The methods used for each of the models are described. The graph searching methods are based on those described by Hart in 1968. The simulated arm is modelled as a line segment fixed at the origin with a skin some constant distance from this line segment. The method for each of the models is described:

(i) **Parallelepiped:** Each obstacle is represented as a parallelepiped by defining the corner points of the obstacle. The obstacles created an obstructed segment of the robot work-space bounded by lines from an apex coincident with the origin. This segment can be simplified to be bounded by four sides and then grown by the radius of the arm. The problem is reduced to a two degree of freedom line moving around these grown segments. This is shown in figure 7.3.

From figure 7.3 it can be seen that using these models, the shortest path consists of planes between these obstacles and the surfaces of particular obstacles. To determine the path a heuristic method of graph searching can be used.

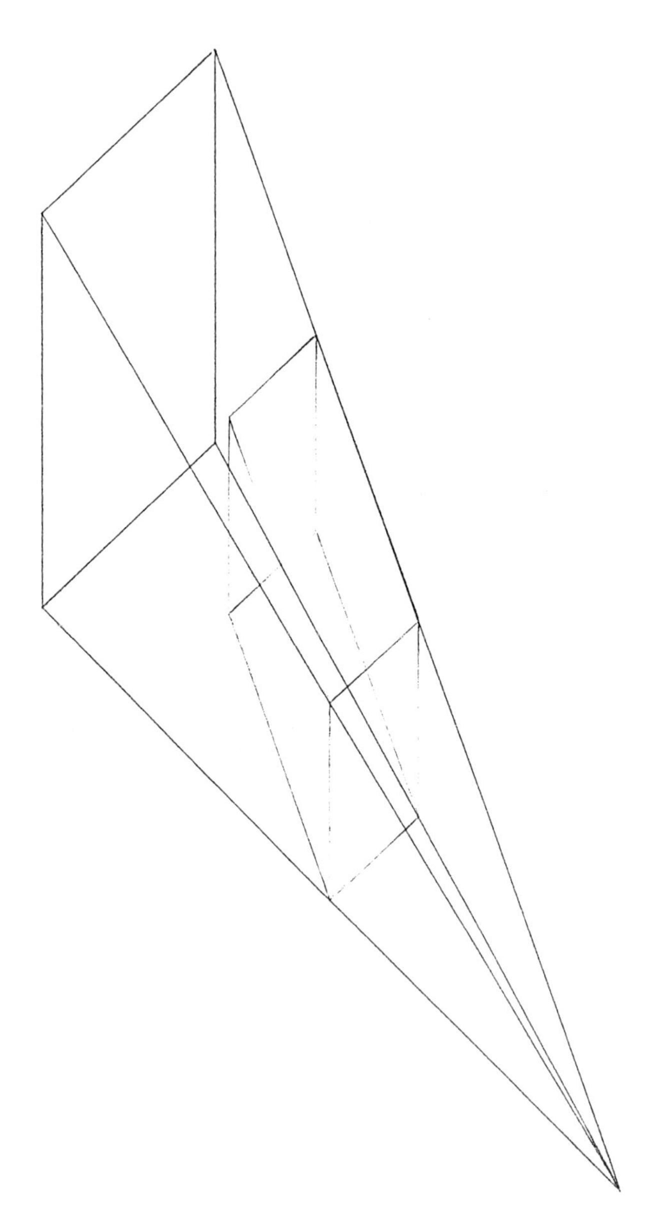

Figure 7.3: 2-D Path Finding Problem around a Parallelepiped.

From the START configuration a graph is generated. Each simplified parallelepiped can be traversed via several edges and faces. Each edge has a node associated with it. In fact only two paths are possible around each obstacle and thus four paths are possible between obstacles. The routine to find the four bounding configurations for the base and simulated shoulder joint which corresponded to edges of the four bounding sides is shown.

```
ShoulderMax =-10 : ShoulderMin =200
BaseMax =-200 : BaseMin =200
FOR Count =1 TO PolyNo
        FOR Corner =1 TO NumberOfCorners[PolyNo]
                Base[Angle] = ATN(Corner[x]/Corner[y])
                FindMod(Corner[x],Corner[y],Mod)
                Shoulder[Angle] =ATN(Corner[z]/Mod)
                IF Base[Angle] >BaseMax THEN
                        BaseMax =Base[Angle]
                ELSE IF Base[Angle] <BaseMin THEN
                        BaseMin =Base[Angle]
                END IF
                IF Shoulder[Angle] >ShoulderMax THEN
                        Shouldermax =Shoulder[Angle]
                ELSE IF Shoulder[Angle] <ShoulderMin THEN
                        ShoulderMin =Shoulder[Angle]
                END IF
        Next Corner
NEXT Count
```

A node is assigned to each of these bounding configurations. From the START configuration each node is tested against a cost function, beginning with the configurations with the lowest base angle. The cost function is defined such that:

$$\text{Cost} = d_{\text{Old-New}} + \Sigma d_{\text{GOAL-New}} - \Sigma d_{\text{GOAL-Old}}$$

where,

$d_{\text{Old-New}}$	=	Distance from old node to new node.
$\Sigma d_{\text{GOAL-New}}$	=	Sum of the distances between nodes from the new node to the GOAL.
$\Sigma d_{\text{GOAL-Old}}$	=	Sum of the distances between nodes from the old node to the GOAL.

Distance is considered to be the total movement of both joints. Later it will be assumed that both joints are capable of similar accelerations and velocities, so that distance is the largest difference. Thus the cost between the old node and the new node is the extra distance that the robot is required to travel plus the new distance to the GOAL compared with the old distance.

(ii) **Spheres:** The obstacles are represented as spheres by defining the centre and radius of the sphere in Cartesian coordinates. The radius is grown by the radius of the robot arm and the problem reduced to a line segment with two degrees of freedom moving among cones. As the robot has two degrees of freedom the spheres formed blocked cones emanating from an apex coincident with the origin. They form circles on the bounding sphere of the robot work area as shown in figure 7.4. From this it can be seen that in this case the shortest path from START to GOAL consists of planes between the cones and arcs around the cones. From the START configuration a graph is generated. Each cone could be traversed in a clockwise or anti-clockwise direction, so each obstacle has only two nodes associated with it. Thus two paths are possible (one to each node) and four paths are possible between cones. An example routine to find the two nodes for a sphere is included on the next page.

A similar cost function is defined for each node so that

$$\textbf{Cost} = d_{\text{Old-New}} + \Sigma d_{\text{GOAL-New}} - \Sigma d_{\text{GOAL-Old}}$$

Thus the cost between the old node and the new node is the extra distance that the robot is required to travel plus the new distance to the GOAL compared with the old distance. The node with the lower cost is selected in each case until the GOAL node is reached.

(b) **A Global Method:** In this case the working area is divided into a discrete graph of joint angles with nodes at increments of 5 degrees, so that the base angles were -180°, -175°, -170°....170°, 175°, 180° and the simulated joint angles were 0°, 5°, 10° ....170°, 175°, 180°.

The method involves testing this discrete graph of 2-space from a START configuration to a GOAL configuration, checking each node for an obstruction. Data about each node is stored in a variable named NodeStatus% in the table in figure 7.6.

An example routine to find the two nodes for a sphere obstacle

```
Value =(ConeRad[No)/(ConeRad[No)+ConeRad[No+1))
PlanePoint(x) =ConeCent(No,x) +Value*CentLine(x)
PlanePoint(y) =ConeCent(No,y) +Value*CentLine(y)
PlanePoint(z) =ConeCent(No,z) +Value*CentLine(z)
ModSquared =(PlanePoint(x)*PlanePoint(x)) +(PlanePoint(y)*PlanePoint(y))
                                    +(PlanePoint(z)*PlanePoint(z))
Modulus =SQR(ModSquared)
DotProd =(PlanePoint(x)*ConeCent(No,x) +PlanePoint(y)*ConeCent(No,y)
                                    +PlanePoint(z)*ConeCent(No,z)) / ModSquared
TempVal =DotProd / ModSquared
PerpPoint(x) =TempVal * ConeCent(No,x)
PerpPoint(y) = TempVal * ConeCent(No,y)
PerpPoint(z) = TempVal * ConeCent(No,z)
FindModulus(PerPoint(),SpherCent(),PerLine(),PerLineModSquared,PerLinMod)
LineNodeModSquared =PerLineModSquared - (R*R)
IF LineNodeModSquared < 0 THEN
        Flag =FALSE   ' Cones are not separated!
ELSE
        LineNodeMod =SQR(LineNodeModSquared)
        Value =LineNodeModSquared/PerLineModSquared
        LineCent(x) =Value * PerLine(x)
        LineCent(y) =Value * PerLine(y)
        LineCent(z) =Value * PerLine(z)
        Value =(LineNodeMod*R)/(PerLineModSquared * Modulus)
        X_Prod(PerLine(),ConeCent(),Return_X())
        Node_a(x) =PerpPoint(x)+LineCent(x)+Value*Return_X(x)
        Node_a(y) =PerpPoint(y)+LineCent(y)+Value*Return_X(y)
        Node_a(z) =PerpPoint(z)+LineCent(z)+Value*Return_X(z)
        Node_b(x) =PerpPoint(x)-LineCent(x)+Value*Return_X(x)
        Node_b(y) =PerpPoint(y)-LineCent(y)+Value*Return_X(y)
        Node_b(z) =PerpPoint(z)-LineCent(z)+Value*Return_X(z)
        X_Prod(PlanePoint(),Node_a(),Return_X())
        Temp =SQR(Return_X(x) * Return_X(x)) +SQR(Return_X(y) * Return_X(y))
                              +SQR(Return_X(z) * Return_X(z))
        Value =     SQR(Temp)
        Node_u(x)  = Return_X(x) / Value
        Node_u(y)  = Return_X(y) / Value
        Node_u(z)  = Return_X(z) / Value
        X_Prod(PlanePoint(),Node_a(),Return_X())
        Temp =SQR(Return_X(x) * Return_X(x)) +SQR(Return_X(y) * Return_X(y))
                              +SQR(Return_X(z) * Return_X(z))
        Value =     SQR(Temp)
        Node_l(x)  = Return_X(x) / Value
        Node_l(y)  = Return_X(y) / Value
        Node_l(z)  = Return_X(z) / Value
END IF
```

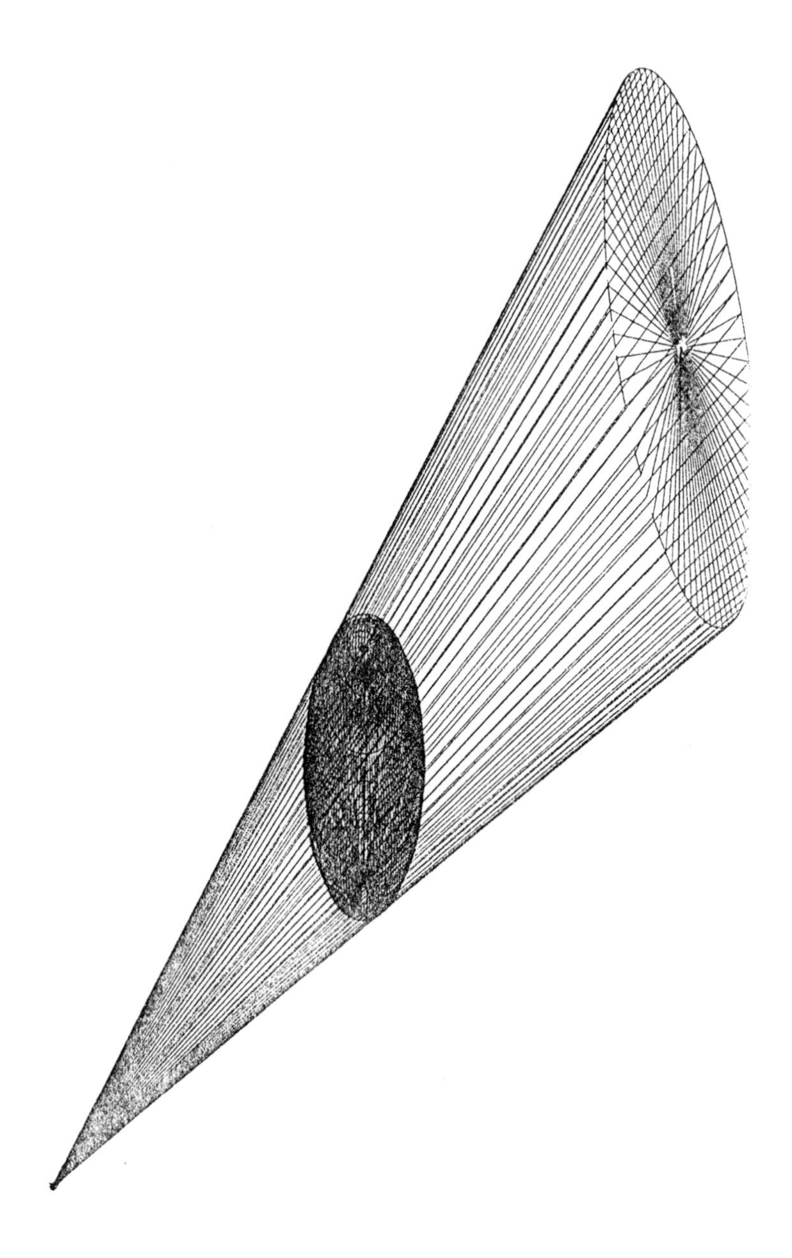

Figure 7.4: 2-D Path Finding Problem through Cones.

The robot will not follow arcs around the circular segments but moves in planes as shown in figure **7.5**. This movement is simpler.

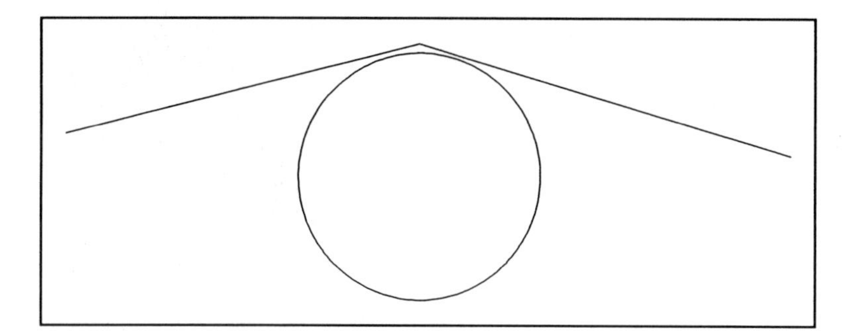

Figure **7.5**: The Planar Movement Around Circular Segments (Viewed from the Origin).

An example program is considered.

When the program starts, an array called NodeStatus% is defined as an array of the graph of joint angles. Each node within the array is set to clear. BLOCKED nodes are loaded from a disc file and bit 6 of these nodes is cleared. With the BLOCKED nodes initialised, the START and GOAL nodes are requested from the operator and the START node is placed onto a list.

The method tests nodes around the graph of joint angles. From the START, each of the four nearest nodes is tested against a cost function to see if they are closer to the GOAL. Any nodes that are nearer are added to the list. The cost function is: $\text{Cost} = (\Sigma d_g)^2 + \Sigma d_s$

where Σd_g = Sum of the distances between nodes from the GOAL.

and Σd_s = Sum of the distances between nodes from the START.

The node stored at the top of the list is the one that takes the least moves to arrive at its present location compared with other nodes equal distances from the GOAL configuration. This node has its nearest three neighbours tested and so on until the test node is the GOAL node.

NodeStatus%		Bit Level	
Bit Value	Bit	0	1
80_H	7	Not On List	On List
40_H	6	BLOCKED	CLEAR
20_H	5	Not Used	Not Used
10_H	4	Not Used	Not Used
08_H	3	Positive Direction	Negative Direction
04_H	2	Base Still	Base Movement
02_H	1	Shoulder Still	Shoulder Movement
01_H	0	Not Used	Not Used

Figure 7.6: Table showing the Bit Assignments for the flag 'NodeStatus%'.

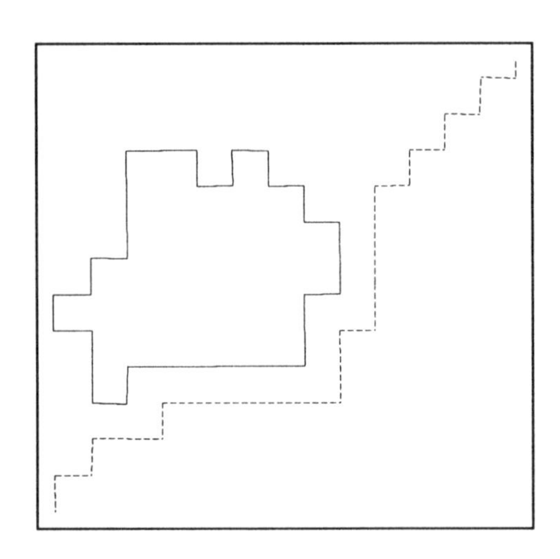

Figure 7.7: A section of a Global Path in 2-D Space.

Each time a node is added to the list, three bits of an array, for example bits 1 to 3 of NodeStatus%, are filled to record the direction moved to arrive at the node. When the test node arrives at the GOAL the path can be retraced by testing these bits. The list can be displayed on a computer screen graphically or stored as data onto a disc. A typical section of a path around a planar simulated complex obstacle is shown in figure 7.7.

7.3 Extension to 3-SPACE Local Heuristic Methods.

In this Section the methods described in 7.2 are extended to plan paths for the three lower joints of a Mitsubishi robot. In 3-D space the simple obstacle models appear as one or more complex shapes in joint space.

For the local and heuristic methods, either of the 2-D planning techniques described in the previous Section are used to plan a path for the lower joints θ_1 and θ_2. The problem is then reduced to finding a path through a new transformed 2-D space for the joint θ_3. This new problem can be solved by a local heuristic method for searching this new 2-D space.

From the 2-D path planning methods a series of configurations of the upper arm are produced. Between these configurations the upper arm moves in planes. The forearm path planning algorithm must avoid obstacles. An initial START configuration and a final GOAL configuration are known. In between these configurations there are configurations where the position of the upper arm is known but the forearm position is undefined.

Sphere models are easily extended to 3-D space but the parallelepiped is complex and requires excessive processing. This is because the sphere models are effectively solid models and only one check is required to see how close the line representing the forearm is to the centre of the sphere. The parallelepiped models are effectively wire frame models defined at their corners so that many calculations are required to see if the forearm violates the obstacle space. For the remainder of this Section sphere models are assumed.

Initially a trajectory locus in which the elbow joint θ_3 moves directly to a final configuration is considered. This path is discretised, only allowing θ_3 to move in multiples of 5 degrees between movements. The positions along this path are checked for collisions. If the path for the forearm is obstructed (as shown in figure 7.8), then a new path is calculated. For the range of configurations through which the forearm moves the sub-range where collision could occur is determined. The configurations at either end of this sub-range are noted. (**A** and **B** in figure 7.8). The range of movement of the base between these points is determined and points are proposed a similar distance above and below the configuration **C**, midway between **A** and **B** on the graph. If one of these configurations is CLEAR, (in this case node **D**), this is adopted as a node and therefore a via-point for the path, otherwise the distance from **C** is increased and the new configurations checked. The new forearm path is then tested at 5 degree intervals and the process repeated if the path is obstructed.

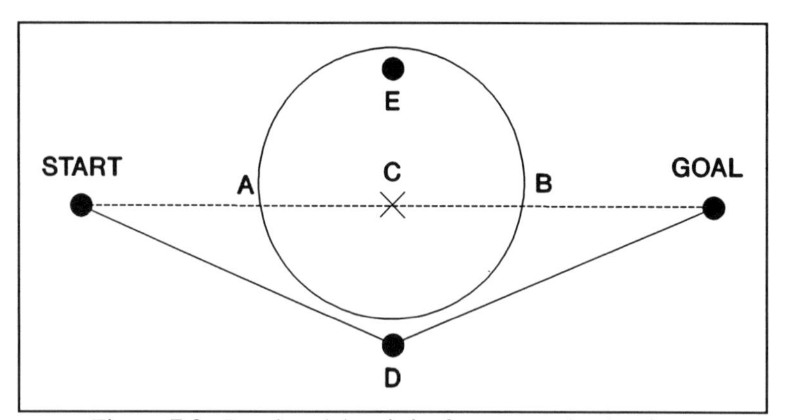

Figure 7.8: Local and heuristic forearm path planning.

7.4 Extension to a 3-SPACE Global Method.

In this Section the 2-D global method described in 7.2 for a robot base and simulated shoulder joint is extended for use with a Mitsubishi robot. The working area is divided into a 3-D graph of joint angles with nodes at increments of 5 degrees, so that:

- The base angles were 30°, 35°, 40°....150°.
- The shoulder angles were -30°, -25°, -20°....110°.
- The elbow angles were 0°, 5°, 10°.... 90°.

Data about each node are stored in the variable, NodeStatus%, which can be extended to include detail on elbow movement as shown in figure 7.9.

NodeStatus%		Bit Level	
Bit Value	Bit	0	1
128 80_H	7	Not On the List	On the List
64 40_H	6	BLOCKED	CLEAR
32 20_H	5	Not Used	Not Used
16 10_H	4	Not Used	Not Used
8 08_H	3	Positive Direction	Negative Direction
4 04_H	2	Base Still	Base Movement
2 02_H	1	Shoulder Still	Shoulder Movement
1 01_H	0	Elbow Still	Elbow Movement

Figure 7.9: Table of the Detail of the Extended Flag 'NodeStatus%'.

When the program begins, NodeStatus% is defined as a 3-D array of the graph of joint angles. Each node within the array is initially set to CLEAR and the BLOCKED nodes are loaded from a data file or received from the vision system. With the graph initialised the START and GOAL nodes can be requested from the operator and the START node is placed onto a list.

From the START configuration each of the six nearest nodes are tested to see if they are closer to the GOAL. The closest node is added to the list. This is repeated at each new node on the graph until the GOAL is reached. The cost function used for the 2-D case is extended for use in a 3-D graph and is

$$\text{Cost} = (\Sigma d_g)^2 + \Sigma d_s$$

where, Σd_g = Sum of the distances between nodes to the GOAL.

Σd_s = Sum of the distances between nodes from the START.

Each time a node is added to the list, bits 0 to 3 of NodeStatus% as shown in figure 7.9 are filled to record the direction moved to arrive at the node. When the test node has arrived at the GOAL the path is retraced using these bits. This list is displayed on the screen.

In the above method and the 2-D method described in Section 7.2 only one joint is moved in each test. This needs to be improved if the movement is to be smooth, the number of nodes in the path reduced and the speed of the movement increased. Two methods are considered:

(i) Test the movement of more than one joint during path planning.

(ii) Find the diagonals within the 3-D graph once a path has been planned.

Processing speed is considered. A diagonal on the global 3-D graph corresponds to more than one joint being in motion. If the diagonals on the graph are tested from a node then as the path is planned there will be a total of 26 tests for each node. The 26 tests are made up of: 6 +12 +8 =26, that is the 6 nodes tested by the previous method, the 12 nodes where two joints are in motion and the 8 nodes where all three joints move. This gives a total of 26 nodes to test each time. These tests are considered for the three possible situations:

(i) If one joint moves in the same direction 3 times then 3x26 =78 tests will take place. 60 more than before.

(ii) If one joint moves in the same direction twice and then another joint moves once, then a two-joint diagonal is in the graph so 52 tests will be computed. 34 more than before.

(iii) If each joint moves once, a three-joint diagonal will have been found with only 26 tests, but this is still 8 more than before.

This method is unattractive and a better method is the processing of the planned path data. A routine is required to scan the path data and modify it to include the diagonal movement of more than one joint. Each joint is tested in turn to find where an angle changed for a second time. In the first example shown, this would be 4 nodes down from the start node, and in the second example it is only 3 nodes down. The node before would be moved up to the start node and the process restarted from that node. In this way the paths shown on the right are produced.

```
60 , 20 , 170          60 , 20 , 170
65 , 20 , 170          65 , 25 , 175
65 , 25 , 175
65 , 25 , 175

60 , 20 , 170          60 , 20 , 170
65 , 20 , 170          65 , 25 , 170
65 , 25 , 170          70 , 25 , 170
70 , 25 , 170
```

The following routine achieves the path modification:

```
'—— include diagonal movement
scanpos% =0
DO
  secchange% =4       ' the offset of the second change in angle
  FOR lp1% =0 TO 2
    numchanges% =0    ' the number of angle changes of the joint
    FOR lp2% =1 TO 3
      diff% =path%(scanpos% +lp2%, lp1%) - path%(scanpos% +lp2% - 1, lp1%)
      IF diff% <>0 THEN         ' is there a change in angle
        numchanges% =numchanges% +1
        IF numchanges% =2 AND lp2% <secchange% THEN secchange% =lp2%
      END IF
    NEXT
  NEXT
  IF secchange% >2 THEN              ' is there a diagonal
    FOR lp1% =scanpos% TO pathpos%   ' move other nodes up
      path%(lp1%, 0) =path%(lp1% +secchange% - 2, 0)
      path%(lp1%, 1) =path%(lp1% +secchange% - 2, 1)
      path%(lp1%, 2) =path%(lp1% +secchange% - 2, 2)
    NEXT
    pathpos% =pathpos% - secchange% +2
  END IF
  scanpos% =scanpos% +1
LOOP UNTIL scanpos% >=pathpos%
```

A second improvement to the path is to remove the repeated nodes due to constant joint motion. This is demonstrated.

40 , -20 , 100 45 , -20 , 100 50 , -20 , 100 55 , -20 , 100	40 , -20 , 100 55 , -20 , 100
60 , 10 , 140 65 , 15 , 140 70 , 20 , 140 75 , 25 , 140 80 , 30 , 140	60 , 10 , 140 80 , 30 , 140

This routine involves finding parts of the path where joint motion is constant and removing all the nodes in between the beginning and end of this movement.

```
'—— remove constant change nodes
path2pos% =1
FOR lp1% =pathpos% - 1 TO 1 STEP -1
  '—— find the change in angle of present and previous nodes
  diff(0)u% =path%(lp1% +1, 0) - path%(lp1%, 0)
  diff(0)d% =path%(lp1%, 0) - path%(lp1% - 1, 0)
  diff(1)u% =path%(lp1% +1, 1) - path%(lp1%, 1)
  diff(1)d% =path%(lp1%, 1) - path%(lp1% - 1, 1)
  diff(2)u% =path%(lp1% +1, 2) - path%(lp1%, 2)
  diff(2)d% =path%(lp1%, 2) - path%(lp1% - 1, 2)                '*** If not the same as
IF NOT (diff(0)u% =diff(0)d% AND diff(1)u% =diff(1)d% AND diff(2)u%=diff(2)d%) THEN
        path2%(path2pos%, 0) =path%(lp1%, 0)                    'previous then store in
        path2%(path2pos%, 1) =path%(lp1%, 1)                    ' other array
        path2%(path2pos%, 2) =path%(lp1%, 2)
        path2pos% =path2pos% +1
  END IF
NEXT
```

Smoothness can now be considered. Using a "string pulling" technique similar to that described by Dupont in 1988 the path can be shortened. With the paths determined above, the movement of the robot is smoother as it leaves the last obstacle and makes for the GOAL than during the rest of the path. This is because the planner is drawn towards the GOAL and has to work around obstacles but could move easily away from the last obstacle. This is shown in figure 7.10. A new path is determined from the configuration midway between the START and the GOAL node configurations, back to the START.

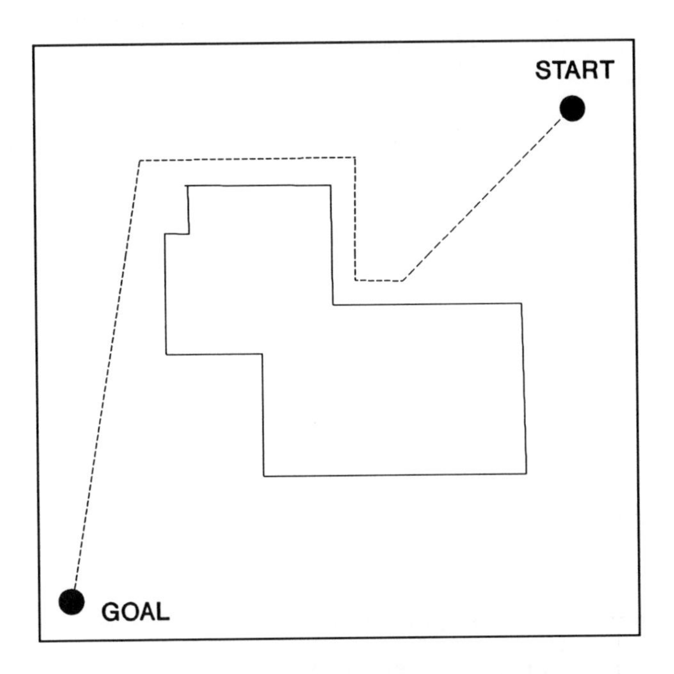

Figure **7.10**: An Initial Path in the Plane.

The original path is stored in an array, the START is re-defined as the GOAL and a new path is determined. The two paths are combined by reversing the new path and adding the end of the original path from the furthest point onwards. This method produces a smoother path as demonstrated for the planar case in figure 7.11.

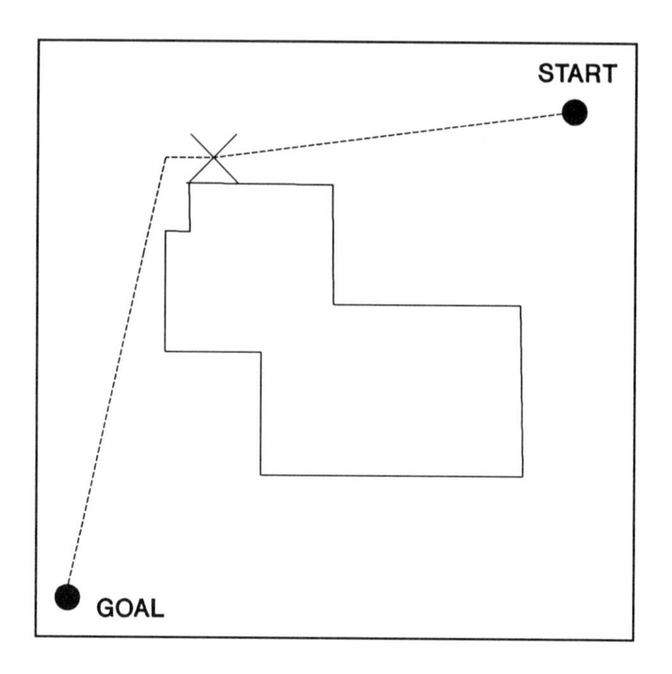

Figure 7.11: Revised Path in the Plane after "String Pulling".

If the direct path from START to GOAL is blocked then the technique in 3-D space scans two lines from the START and GOAL, travelling along the joint in which there is most difference. The code is shown below:

```
FUNCTION ScanLine1% (s1%, s2%, s3%, e1%, e2%, e3%)
hit% =0
FOR lp% =s1% TO e1%
  IF NodeStatus%(lp%, CINT(s2% +(lp% - s1%) * (e2% - s2%) / (e1% - s1%)), CINT(s3% +(lp%
- s1%) * (e3% - s3%) / (e1% - s1%))) AND hittarget <>0 THEN hit% =1
NEXT
ScanLine1% =hit%
END FUNCTION
```

A revised technique produces via-points to smooth the path and increase the speed of the trajectory produced. The joint with most difference is selected. For a pick and place robot this is often the base of the robot so this is used in the description; a line is tested from the START with the shoulder and elbow staying at the angles of the START configuration. Another line is scanned in the same manner but from the GOAL. In this way the range where a collision would occur is found. This is similar to points **A** and **B** in figure 7.8. The node on the original path which is a maximum distance away from the START configuration is used to define the angles for joints θ_2 and θ_3 for two via-points. The base angles $\theta_{1v(i)}$ and $\theta_{1v(ii)}$ for these via-points are the base angles at the extremes of the range where a collision would have occurred. A line is scanned from the START to GOAL of the obstacle at the shoulder and elbow angles of the furthest point, that is between the two proposed via-points. If this line is obstructed then the shoulder and elbow angles are moved out until a clear path is found. The whole new path including the two new via-points is then tested, and if clear, the path is passed to the Robot Controller.

For the simple obstacle models considered in this book this path contains four nodes and the method works in all the practical situations tested by the author.

7.5 Trajectory Generation.

Once a path is planned as a trajectory locus in joint space, the configurations are passed to a controller to generate the motions. In the system described earlier this is the control computer. Although the path is discretised, the intermediate configurations are close to each other so that the trajectory results in a curve in joint space that is close to the planned trajectory locus.

The path can be transferred using a floppy disc or by a serial link, such as the RS 232 link described in Chapter three. The path is a set of joint angles in the order in which they must be moved. The path movement routines in the controller typically move each joint to within 8 encoder readings or 0.1° for the configurations specified at via-points and to within 2 encoder readings or 0.025° at the GOAL configuration.

The planning routines described in this book were demonstrated at Portsmouth Polytechnic in 1989 using a Mitsubishi robot and a simple environment. The robot controller carried out a simulated task in the waist range +30°to -30° until a path was received from the main computer. Then the robot moved along the path to the destination and returned to the simulated task until a new path was received.

The robot trajectory locus consisted of a series of robot configurations in joint space. The robot control computer also operated in joint space but the controller did not always interpolate between coordinate positions in a predictable manner. To avoid this problem, intermediate configurations were generated by the path planner as robot coordinates. This meant the robot moved only small distances between defined configurations and hence the deviation from the path was negligible.

7.6 Results.

In earlier Chapters, polyhedral models were selected for the static environment in the global path planner. The BLOCKED nodes due to the polyhedral models were calculated once and stored in the main computer. These nodes are loaded at the beginning of each planning session.

When the camera was introduced the static environment was simplified to a single polyhedron modelling the floor of a work-place (The top of a laboratory work bench). The positions of the ForeTip for the BLOCKED configurations are shown below in figure 7.12.

The model of the static environment reduces the volume of discrete space left available to the planner. The position of the ForeTip for the remaining CLEAR configurations are shown in figure 7.13.

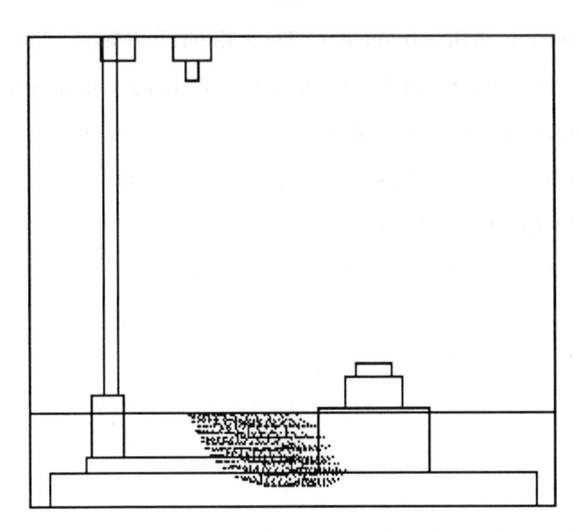

Figure 7.12: The BLOCKED nodes stored for the static environment.

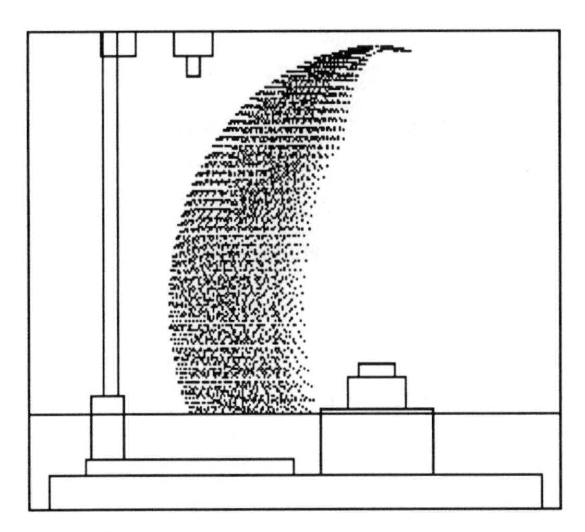

Figure 7.13: The configurations available without obstacles.

The space available to the path planner is further reduced by the introduction of obstacles into the work place. Figures 7.14 to 7.16 show the position of the ForeTip for the blocked configurations due to two types of obstacle, a cube and a cylinder (a beer can spray-painted black). The 2-D slice model is used in all cases.

The phrase "real-time" has several interpretations and in this book it is assumed to mean that the solution of the path planning problem takes less time than the machinery takes to execute the motions.

The total processing time for the system depended on:

(a) The number of obstacles.
(b) The position of the obstacles.
(c) The size of the obstacles.

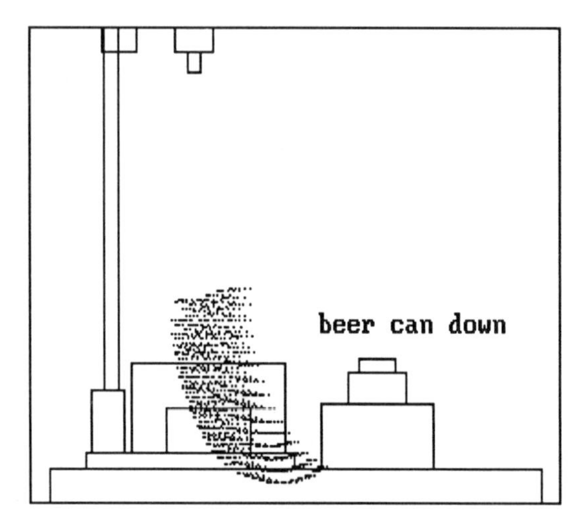

Figure 7.14: Blocked Configurations due to the Horizontal Cylinder.

In practice solutions to the path planning problem are always found by both methods and the calculation time is within the limits for this definition of real-time operation. For a typical task, such as:

> Simulate a task between base angles +30° and -30°, then "pick up" a part at one extreme of the area covered by the vision system, (base angle +30°) and move it to the other extreme (+150°), while avoiding an obstacle.

The robot trajectory tended to take >9seconds and the total calculation time after inserting an obstacle was <9seconds. The internal timer was interrogated during path processing. The local and heuristic 3-D space path planner tended to produce paths using the sphere representation within 3 seconds. The global path planner produced paths using 2-D slices in joint space within 1.4 seconds.

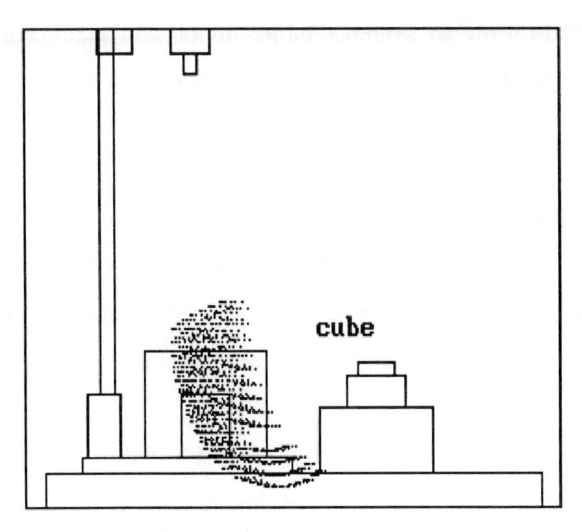

Figure 7.15: Blocked Configurations due to the Cube.

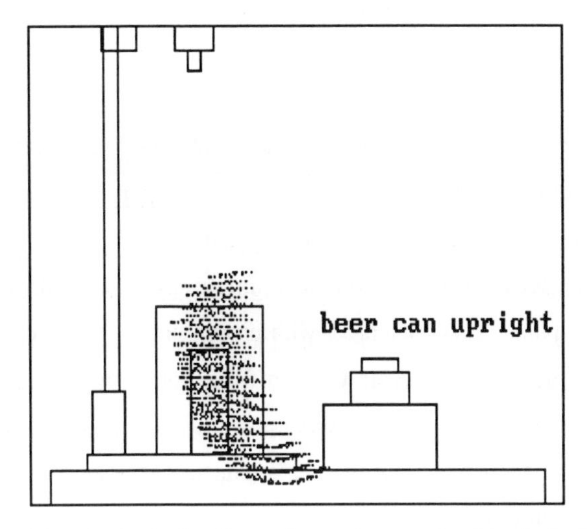

Figure 7.16: Blocked Configurations due to the Vertical Cylinder.

To show some typical paths planned by the robot within this book, the robot was simulated using the forward kinematic solutions for the robot described in Chapter three. The graphics facilities of Quick BASIC were used to draw the robot on the screen and this was captured using the GRAB feature of the Word Perfect word processor. The simulation methods were those described by Moore in 1990.

Figures 7.17 to 7.24 show an example of the robot arm moving along a planned path around a point object at X=0, Y=210, Z=50.

7.7 Discussion and Conclusions.

To achieve the simulation displayed in figures 7.17 to 7.24, the basic forward kinematics described in Chapter three were adapted to the (X,Y) coordinates of the screen. This was achieved using the two lines of code describing the kinematics of the robot:

```
LINE (x,y) - (x-sf*SIN(Θ1)*220*COS(Θ2), y - sf*220*SIN(Θ2))

LINE (x-sf*SIN(Θ1)*220*COS(Θ2), y - sf*220*SIN(Θ2)) -
          (x-sf*SIN(Θ1)*(220*COS(Θ2)+160*COS(Θ2 +Θ3 - π)),
          y - sf*(220*SIN(Θ2)+160*SIN(Θ2 +Θ3 - π))
```

To this solution is added the skin of the robot. The path shown used the global path planning routines which produced paths that incremented joints by 5° per move.

The calculation time for both the planning methods is adequate, but the local and heuristic methods tend to take twice as long compared to the global method. This is partly due to the calculations for the static environment being calculated every time for the local method. For the example, both methods took less than 5 seconds to plan a path. This compared with programming times of 5 to 20 minutes for programmers using the GRASP CAD off-line robot programming package.

In practical environments the path planning computer always produced satisfactory paths in "real time" but the performance of the automatic programming system could be improved by:

(a) A Cartesian robot which would simplify the algorithms.

(b) Parallel processing computers.

(c) Improvements in software techniques.

(d) Faster processing speeds.

Figure 7.17

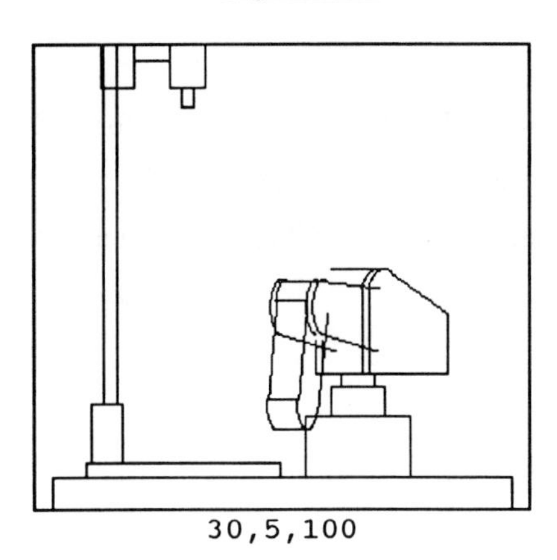

30,5,100

Figure 7.18

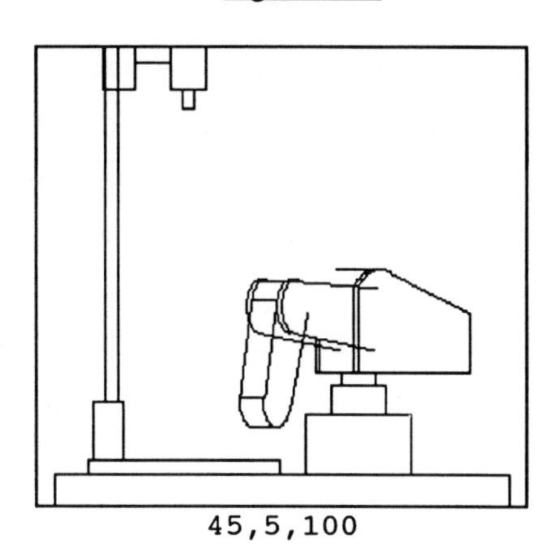

45,5,100

Figure 7.19

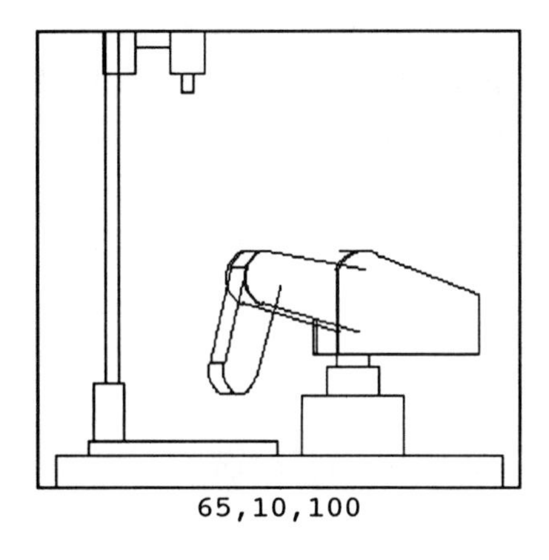

65,10,100

Figure 7.20

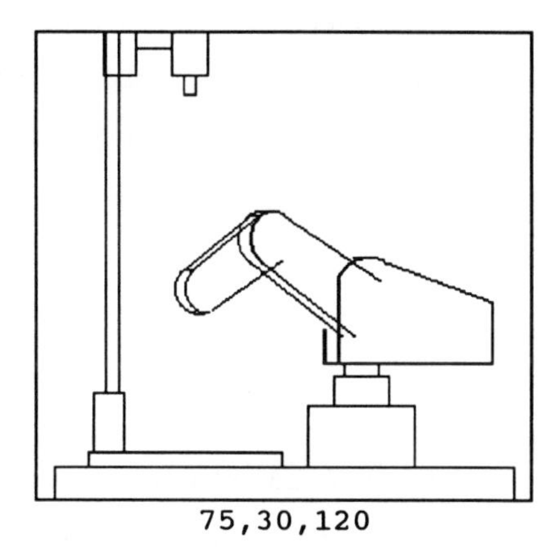

75,30,120

Figure 7.21

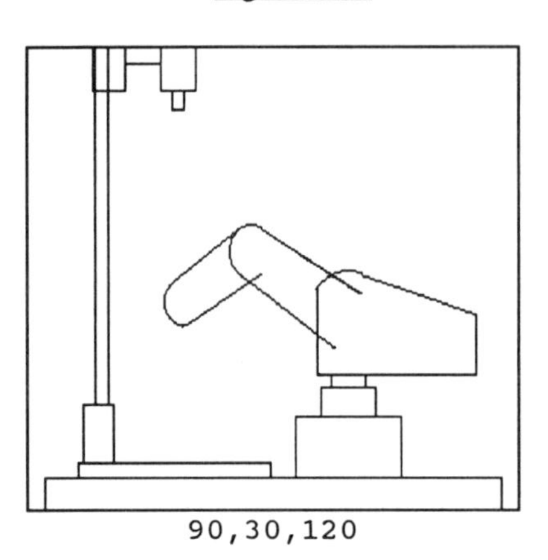

90,30,120

Figure 7.22

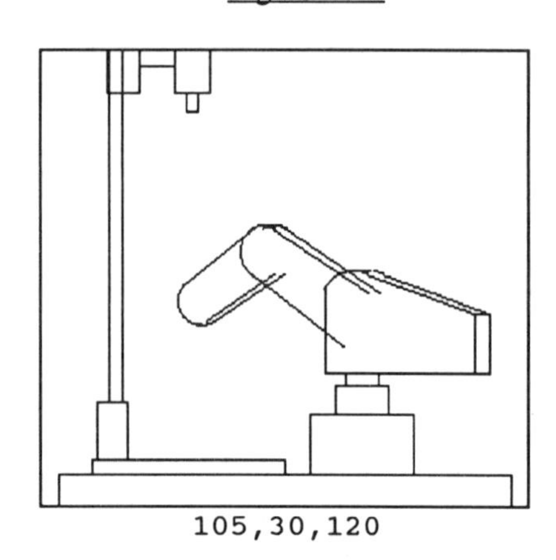

105,30,120

Figure 7.23

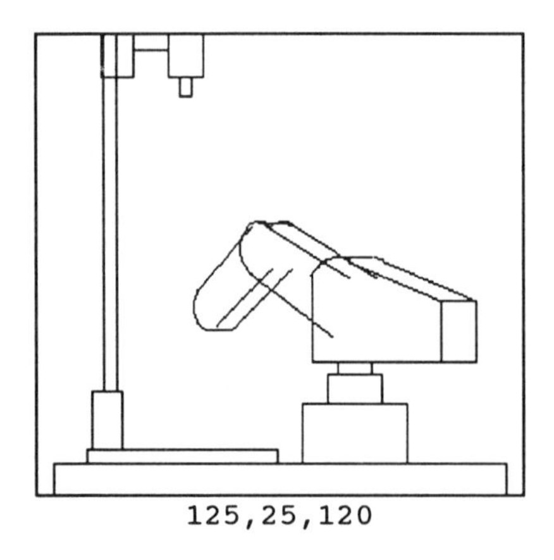

125,25,120

Figure 7.24

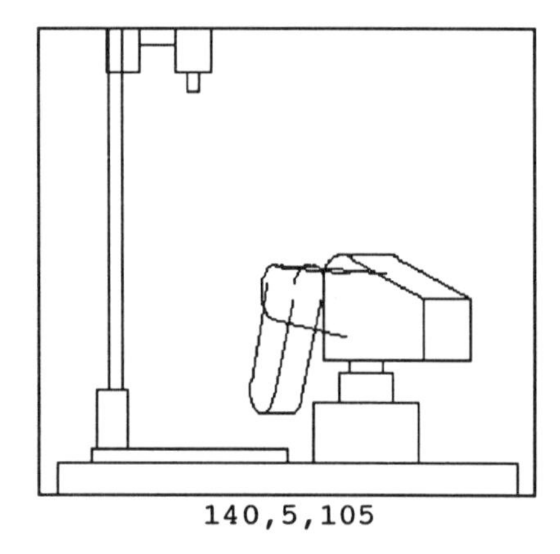

140,5,105

The advantage of the methods described are that they use simple rules to solve problems which are difficult to analyze. The disadvantages of the local and heuristic methods are that they may not always find a path where one exists and the static environment is considered for every path. The fact that the forearm and upper arm are planned separately means that many possible solutions are not considered and hence the paths produced are unlikely to be the best path.

The local and heuristic programs are closely tied to the configuration of the robot. Some of the program code would need to be modified to accommodate the kinematic chain of a different robot. The 3-D global method is a more general solution to the problem and changing the robot would just require changing a module in a program.

For the global methods, the blocked nodes for the 3-D graph are loaded from disc or received from the vision system. The transformations are described in earlier Chapters for the 2-D slice models and the sphere models. The 2-D slice models give the best performance.

For the local heuristic methods the transformations took place during path processing and although the parallelepiped models performed better than the sphere models in 2-D space, in 3-D space, the sphere models were more efficient. This is due to the nature of the stored data concerning the obstacles, in that the sphere model is effectively a solid model while the parallelepiped is effectively a wire frame model.

Chapter Eight

PATH IMPROVEMENT CONSIDERING THE MANIPULATOR DYNAMICS

8.1 Introduction.

This chapter will explore methods of adapting robot paths to produce faster and more efficient robot trajectories.

Existing methods for programming robots and complex machines generate paths which may appear simple or obvious to the operator but which may not be efficient for the robot. Once a robot has been programmed to work within a complex system, possibly without the programmer ever seeing the work-place, it may be possible to improve the solution. This can provide the robot with a degree of autonomy.

Recent commercial robotic CAD systems allow dynamic modelling of robots and machine tools within flexible assembly systems. Cell lay-out can be improved by testing various configurations and running different robot programs to optimise the cell design and the product construction sequence. The programs produced can then be used through post-processors to directly program the robots on the factory floor.

Within computer design systems, complicated functions of space and time are decoupled from the operator and only simple descriptions of the desired motion are considered. The paths produced pass through "via-points" where joint velocities may change abruptly.

Positions calculated by off-line computer programming or CAD post-processors are usually represented in a coordinate frame (usually Cartesian space) related to the joint variables by some homogeneous transform. In CAD systems, objective level programming tends to be used and this was reported by Snyder in 1975. The objective level program relates end effector position to a work-piece or object in the cell. This type of programming is easier for operators to visualise and model. Motions of the manipulator are described as motions of the tool frame relative to the world frame. Little consideration is given to the dynamics of the robot.

Once programmed with a set of space and time coordinates, a simple robot will carry out a sequence of motions with little sensing of the environment and with little correction once set in motion. The end effector paths produced by CADCAM {or off-line programming} may be "too specific" and therefore the joint trajectories more complex than is required for these simple tasks. For first generation robot tasks, such precision is not always necessary. Other methods, such as teaching by following, produce a continuous path control that appears simple and ordinary to the human operator, but which may generate via-points which cause unnecessary current transients and torques in the electrical actuators. Motions may be further complicated by physical or safety restrictions for human teachers within the robots working volume.

Chapters three to six have described the creation of an automatic path planner in order to increase productivity. This chapter presents a method to fulfil a further aim: To produce systems which would improve the performance of robots for which paths had already been planned by some means, automatic or otherwise.

Robots and machines are physical systems and are subject to physical limitations. By considering these limitations, performance can be improved with reference to some criterion, and refined motions calculated for robots and complex machines.

A typical robot task is selected as an example, a repetitive series of pick and place movements. In this situation the automatic reprogrammer can take some time in calculating the improved paths while the robot carries out its original program. The motions can be modified when the set of destinations is repeated.

The method of path improvement presented in this chapter uses a simple model of the robot dynamics to improve a given task.

Models of the dynamics for active mechanisms are complex and many procedures for generating models have been devised; some are described by Brady et al in their book from 1982 and a dynamics model for a manipulator carrying loads was derived by Izaguirre & Paul in 1985.

Two major approaches in terms of the formulation of robot dynamics equations are the Newton-Euler method and the Lagrangian formulation. The Newton-Euler method solves the problem recursively to find joint torques one by one whereas the Lagrangian method solves the problem using closed-form differential equations.

An, Atkeson & Hollerbach in 1986 employed the Newton-Euler formulation to determine the inertial parameters of robot links. These could then be used in the recursive dynamics computation described by Fu, Gonzalez and Lee in their book from 1987. In 1985 Neumann & Khosla adopted a hybrid procedure combining the Newton-Euler and Lagrange formulation of the dynamics to estimate the inertial parameters of more than one link. The Lagrangian formulation was first developed to compute closed-form manipulator dynamics by Uicker in 1966 and later in 1969 by Kahn.

In 1985 Mukerjee & Ballard used full torque sensing at each joint of a complex machine to determine the link parameters and establish a tabular friction model. Haddad in 1985 and later Kumar in 1988 employed the Lagrange formulation for the case of a manipulator with two rotary joints. Olson & Bekey used joint torque sensing in 1985 during single joint motion to estimate the link parameters for rotary joints.

Even though many of the theoretical problems in manipulator dynamics have been solved, the question of how to best apply the theories to robot manipulators is still being debated. In the work presented in this Chapter, information on system dynamics is used to produce a set of simple rules for an automatic motion improvement system.

The dynamics of the manipulator in closed form Lagrange equations were selected to represent the dynamics by a set of second-order coupled non-linear differential equations. The form of these equations was exploited in an attempt to establish a set of simple rules. An experimental procedure was applied to the Mitsubishi RM 501 robot described in Chapter three. The measured quantities were the drive currents to the motors (which represented the torques) and the joint angular positions. This method is similar to the methods described by Kumar for a two-link planar robot manipulator in 1988. The advantage of using this input-output form is that intermediate non-linearities (such as gear friction) and the motor characteristics are directly incorporated into the model. The results are unexpected and the model of the robot dynamics is discussed in Section 8.6.

In the next Section the Lagrange formulation for a manipulator with three joints is outlined. The mathematics included in Section 8.2 are involved, and the reader may wish to skip to Section 8.3. In Sections 8.3 and 8.4 the experimental identification procedure is described and in Section 8.5 the interesting results from the application of this procedure are presented. Section 8.6 describes the simple rules developed from these results and Section 8.7 presents the results of using these rules. The Chapter concludes with discussion and conclusions in Section 8.8.

8.2 The Lagrangian Formulation for a Manipulator with Three Joints.

The formulation was based on the Lagrangian equation in terms of the Lagrangian coordinates q given by:

$$\tau_i = \frac{d}{dt}\frac{\partial L}{\partial(dq_i/dt)} - \frac{\partial L}{\partial q_i}$$

where,

L = The Lagrangian function.

q_i = The coordinate of the ith element used to express the kinetic and potential energies.

τ_i = The torque.

The relationships between the torques and the angular positions, velocities and accelerations of the links are obtained by considering the potential and kinetic energies. The Lagrangian L is defined as the difference between the kinetic and potential energy given by: L = **K** - **P**

where:

K is the total kinetic energy.

P is the total potential energy.

In this Chapter, using the expressions for **K** and **P** in terms of manipulator parameters, the equations for the dynamics of the three main links of the Mitsubishi robot are obtained in the form:

$$\tau_i = \sum_{j=1}^{N} J_{ij} d^2\theta_j/dt + \sum_{j=1}^{N} \sum_{k=1}^{N} H_{ijk}(d\theta_j/dt)(d\theta_k/dt) + G_i$$

A Mitsubishi robot was assumed to consist of two main movable links; L_1 and L_2 of masses m_1 and m_2 which could be rotated through angles θ_2 and θ_3, as shown in sketch form in figure 8.1. The robot base L_0, with mass m_0 could rotate through θ_1. To determine the total kinetic and potential energy for the robot, each link is considered in turn.

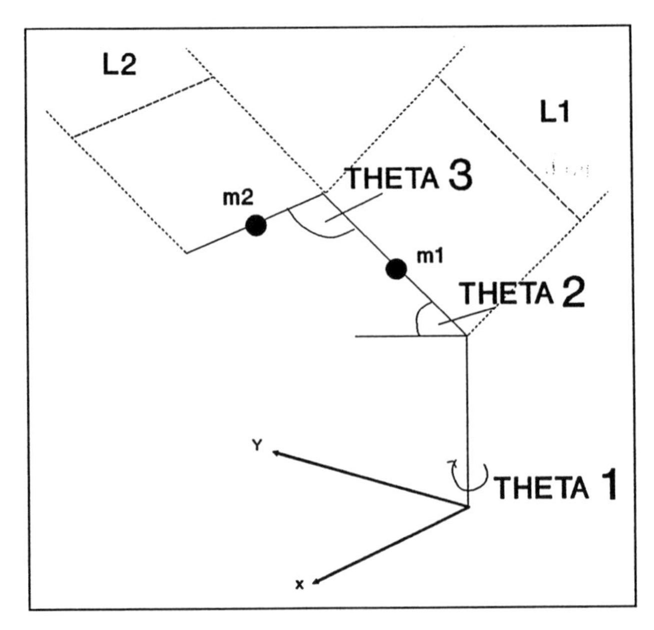

Figure 8.1: The model used for the three main links of the Mitsubishi robot.

The kinetic energy and potential energy equations of link L_0 were assumed to be: $K_0 = I(d\theta_1/dt)^2$

$P_0 = 0$

where I is the moment of inertia of link L_0 about the Z axis.

Considering link L_1, expressing the Cartesian coordinates of the assumed centre of mass shown in figure 8.1 in terms of the joint angles gave:

$$X_1 = L_1/2 \cos\theta_1 \cos\theta_2$$

$$Y_1 = L_1/2 \sin\theta_1 \cos\theta_2$$

$$Z_1 = L_0 + L_1/2 \sin\theta_2$$

Taking derivatives of the equations with respect to time gives:

$$dX_1/dt = -L_1/2\, d\theta_1/dt \sin\theta_1 \cos\theta_2 - L_1/2\, d\theta_2/dt \cos\theta_1 \sin\theta_2$$

$$dY_1/dt = L_1/2\, d\theta_1/dt \cos\theta_1 \cos\theta_2 - L_1/2\, d\theta_2/dt \sin\theta_1 \cos\theta_2$$

$$dX_1/dt = L_1/2\, d\theta_2/dt \cos\theta_2$$

Considering V_1^2 where $V_1^2 = (dX_1/dt)^2 + (dY_1/dt)^2 + (dZ_1/dt)^2$

Using trigonometric identities to reduce the solution, the square of the velocity vector is: $V_1^2 = (L_1/2)^2 (d\theta_2/dt)^2 + (L_1/2)^2 (d\theta_1/dt)^2 \cos^2\theta_2$

The kinetic energy term and the potential energy term of link L_1 are thus assumed to be:

$$K_1 = 1/2\, m_1 V_1^2$$

$$= 1/2\, m_1 (L_1/2)^2 \{(d\theta_2/dt)^2 + (d\theta_1/dt)^2 \cos^2\theta_2\}$$

and

$$P_1 = m_1 g L_0 + m_1 g (L_1/2) \sin\theta_2$$

where g = gravitational acceleration.

The Cartesian coordinates of the centre of mass of link L_2 are assumed to be:

$$X_2 = L_1 \cos\theta_1 \cos\theta_2 + L_2/2 \cos\theta_1 \cos(\theta_2 + \theta_3 - \pi)$$

$$Y_2 = L_1 \sin\theta_1 \cos\theta_2 + L_2/2 \sin\theta_1 \cos(\theta_2 + \theta_3 - \pi)$$

$$Z_2 = L_0 + L_1 \sin\theta_2 + L_2/2 \sin(\theta_2 + \theta_3 - \pi)$$

Taking derivatives of the equations with respect to time:

$$dX_2/dt = -d\theta_1/dt \{L_1 \sin\theta_1 \cos\theta_2 + (L_2/2) \sin\theta_1 \cos(\theta_2 + \theta_3 - \pi)\}$$

$$- d\theta_2/dt \{L_1 \cos\theta_1 \sin\theta_2 + (L_2/2) \cos\theta_1 \sin(\theta_2 + \theta_3 - \pi)\}$$

$$- d\theta_3/dt \{(L_2/2) \cos\theta_1 \sin(\theta_2 + \theta_3 - \pi)\}$$

$$dY_2/dt = d\theta_1/dt\ \{L_1\cos\theta_1\cos\theta_2 + (L_2/2)\cos\theta_1\cos(\theta_2+\theta_3-\pi)\}$$
$$- d\theta_2/dt\ \{L_1\sin\theta_1\sin\theta_2 + (L_2/2)\sin\theta_1\sin(\theta_2+\theta_3-\pi)\}$$
$$- d\theta_3/dt\ \{(L_2/2)\sin\theta_1\sin(\theta_2+\theta_3-\pi)\}$$

$$dX_2/dt = d\theta_2/dt\ \{L_1\cos\theta_2 + (L_2/2)\cos(\theta_2+\theta_3-\pi)\}$$
$$+ d\theta_3/dt\ \{(L_2/2)\cos(\theta_2+\theta_3-\pi)\}$$

So that after reducing the solution using trigonometric identities, the expression for the square of the velocity vector is:

$$V_2^2 = (dX_2/dt)^2 + (dY_2/dt)^2 + (dZ_2/dt)^2$$
$$= (L_1^2+L_2^2/4)(d\theta_2/dt)^2 + L_1L_2(d\theta_2/dt)^2\cos(\theta_2+\theta_3-\pi)$$
$$+ L_2^2/4(d\theta_3/dt)^2 + L_2^2/4(d\theta_2/dt)(d\theta_3/dt)\sin^2(\theta_2+\theta_3-\pi) +$$
$$(d\theta_1/dt)^2\ \{(L_2^2/4)\cos^2(\theta_2+\theta_3-\pi) + L_1^2\cos^2\theta_2 + L_1L_2\cos\theta_2\cos(\theta_2+\theta_3-\pi)$$

and the kinetic energy and potential energy terms are therefore given by:

$$K_2 = (m_2/2)(L_1^2+L_2^2/4)(d\theta_2/dt)^2 + (m_2/2)L_1L_2(d\theta_2/dt)^2\cos(\theta_2+\theta_3-\pi)$$
$$+ m_2L_2^2/8(d\theta_3/dt)^2 + (m_2L_2^2/8)(d\theta_2/dt)(d\theta_3/dt)\sin^2(\theta_2+\theta_3-\pi)$$
$$+ (d\theta_1/dt)^2(m_2/2)\{(L_2^2/4)\cos^2(\theta_2+\theta_3-\pi) + L_1^2\cos^2\theta_2$$
$$+ L_1L_2\cos\theta_2\cos(\theta_2+\theta_3-\pi)$$

$$P_2 = m_2gL_0 + m_2gL_1\sin\theta_2 + m_2g(L_2/2)\cos(\theta_2+\theta_3-\pi)$$

Having found the kinetic and potential energies for the three joints, the Lagrangian of the robot:

$$\mathbf{L} = K_0 + K_1 + K_2 - (P_0 + P_1 + P_2)$$

was calculated so that:

$$\mathbf{L} = I(d\theta_1/dt)^2/2 + 1/2\ m_1(L1/2)^2\ \{(d\theta_2/dt)^2 + (d\theta_1/dt)^2\cos^2\theta_2\}$$
$$+ (m_2/2)(L_1^2+L_2^2/4)(d\theta_2/dt)^2 + (m_2/2)L_1L_2(d\theta_2/dt)^2\cos(\theta_2+\theta_3-\pi)$$
$$+ m_2L_2^2/8(d\theta_3/dt)^2 + (m_2L_2^2/8)(d\theta_2/dt)(d\theta_3/dt)\sin^2(\theta_2+\theta_3-\pi) +$$
$$(d\theta_1/dt)^2m_2/2\{(L_2^2/4)\cos^2(\theta_2+\theta_3-\pi) + L_1^2\cos\theta_2 + L_1L_2\cos\theta_2\cos(\theta_2+\theta_3-\pi)\}$$
$$- m_1gL_0 - m_1g(L_1/2)\sin\theta_2 - m_2gL_0 - m_2gL_1\sin\theta_2 - m_2g(L_2/2)\cos(\theta_2+\theta_3-\pi)$$

The following six derivatives can then be found, $\partial L/\partial\theta_1$, $\partial L/\partial\theta_2$, $\partial L/\partial\theta_3$, $\partial L/\partial(d\theta_1/dt)$, $\partial L/\partial(d\theta_2/dt)$ and $\partial L/\partial(d\theta_3/dt)$ so that the Lagrangian equation in terms of the robot joints:

$$\tau_i = \frac{d}{dt} \frac{\partial L}{\partial(d\theta_i/dt)\ \partial} - \frac{\partial L}{\theta_i}$$

can be applied for each of the links θ_1, θ_2 and θ_3 in turn.

The first dynamics equation is thus:

$$\tau_1 = (d^2\theta_1/dt^2).\{I + m_1(L_1/2)^2\sin^2\theta_2 + m_2(L_1/2)\sin\theta_2 + m_2(L_2/2)\sin\theta_3\}$$

$$+d\theta_1/dt\ d\theta_2/dt\ \ 2.\{m_1(L_1/2)\cos\theta_2 - m_2L_1{}^2\cos\theta_2\sin\theta_2$$

$$+m_2L_1(L_2/2)\cos\theta_2\cos\theta_3\}$$

$$+d\theta_1/dt\ d\theta_3/dt\ \ 2.\{m_2(L_2/2)^2\cos\theta_2\cos\theta_3 + m_2L_1(L_2/2)\sin\theta_2\sin\theta_3\}$$

This equation and the other torque equations had several components. They were:

- Effective inertias (and coupling inertias).
- Coriolis and centripetal coefficients.
- Gravity loadings.

so the equation for τ_1 can be expressed in the form:

$$\tau_1 = D_{11}\ d^2\theta_1/dt^2 + D_{12}\ d\theta_1/dt\ d\theta_2/dt + D_{13}\ d\theta_1/dt\ d\theta_3/dt + D_{1g}$$

where:

D_{11}	=	The effective moment of inertia about the Z1 axis
$D_{12}\ d\theta_1/dt\ d\theta_2/dt$	=	The Coriolis torque acting at joint θ_1 due to the velocities of the base θ_1 and shoulder θ_2.
$D_{13}\ d\theta_1/dt\ d\theta_3/dt$	=	The Coriolis torque acting at joint θ_1 due to the velocities of the base θ_1 and the elbow θ_3.
D_{1g}	=	The gravitational torque.

The second dynamic equation is:

$$\tau_2 = d^2\theta_2/dt^2 \{m_1(L_1/2)^2 + m_2L_2^{\,2}\} + d^2\theta_3/dt^2 \{m_2L_1(L_2/2)\sin(\theta_2+\theta_3-\pi)\}$$

$$+ d\theta_2/dt\ d\theta_3/dt\ 2m_2L_1(L_2/2)\cos(\theta_2+\theta_3-\pi)$$

$$- (d\theta_3/dt)^2\ 2\{m_2L_1(L_2/2)\cos(\theta_2+\theta_3-\pi)\}$$

$$- (d\theta_1/dt)^2 \{m_1(L_1/2)^2\cos\theta_2\sin\theta_2 + m_2L_1^{\,2}\cos\theta_2\sin\theta_2$$

$$+ m_2L_2(L_1/2)\cos\theta_2\cos\theta_3\}$$

$$- m_1g(L_1/2)\cos\theta_2 - m_2gL_1\cos\theta_2 - m_2g(L_2/2)\cos(\theta_2+\theta_3)$$

where τ_2 is the torque applied to θ_2. This equation in coefficient form is:

$$\tau_2 = D_{2I}\ d^2\theta_2/dt^2 + D_{22}\ d\theta_2/dt\ d\theta_3/dt + D_{2cI}\ d^2\theta_3/dt^2$$

$$+ D_{24}\ (d\theta_3/dt)^2 + D_{25}\ (d\theta_1/dt)^2 + D_{26}$$

where

D_{2I}	=	The effective moment of inertia about the Z_2 axis
$D_{22}\ d\theta_2/dt\ d\theta_3/dt$	=	Coriolis torque due to velocities of the shoulder and elbow.
D_{2cI}	=	Coupling inertia term between links L_1 and L_2.
$D_{24}\ (d\theta_3/dt)^2$	=	Centripetal torque at θ_2 due to the velocity of θ_3.
$D_{25}\ (d\theta_1/dt)^2$	=	Centripetal torque at θ_2 due to the velocity of θ_1.
D_{2g}	=	The gravitational torque.

The third dynamics equation is:

$$\tau_3 = d^2\theta_3/dt^2 m_3(L_2/2)^2 + d^2\theta_2/dt^2[m_3L_1(L_2/2)\sin(\theta_2+\theta_3-\pi)]$$

$$+ (d\theta_1/dt)^2\{m_3(L_2/2)^2\sin\theta_3 + m_3L_1(L_2/2)\sin\theta_2\sin\theta_3\ \}$$

$$+ (d\theta_2/dt)^2\{m_3L_1(L_2/2)\cos(\theta_2+\theta_3-\pi)\} - m_2g(L_2/2)\cos(\theta_2+\theta_3-\pi)$$

and in the coefficient form,

$$\tau_3 = D_{3I}d^2\theta_3/dt^2 + D_{3cI}d^2\theta_2/dt^2 + D_{33}(d\theta_1/dt)^2 + D_{34}(d\theta_2/dt)^2 + D_{3g}$$

where:

D_{3I} = The effective inertia term at joint 3.

D_{3cI} = The coupling inertia term between links L_1 and L_2.

$D_{33}(d\theta_1/dt)^2$ = Centripetal torque acting at θ_3 due to velocity $d\theta_1/dt$.

$D_{34}(d\theta_2/dt)^2$ = Centripetal torque acting at θ_3 due to velocity $d\theta_2/dt$.

D_{3g} = The gravitational torque.

The expressions for the dynamics derived in this Section consists of variables, which are functions of sines and cosines of joint positions and constants which depend on manipulator link parameters such as link mass, centre of mass, and radii of gyration. Measurements can be taken of the links to obtain the dimensions of centres of mass and radius of gyration for each link. The link masses can be calculated from the measurements and the density of the materials and then the dynamics constants calculated.

Although values might be calculated from measurements and drawings, the process is tedious. Measurement of parameters such as location of centre of masses and exact shapes is susceptible to errors. An alternative approach is to obtain the constants by actually running the manipulator. This approach uses direct input-output measurements during actual motion and then uses results, such as those presented in Section 8.4, to produce simple rules for robot path improvement. In the next Section an experimental method for determining the coefficients is discussed.

8.3 The Formulation of Experiments to Determine the Dynamic Model.

In 1974 Bejczy first noticed the disparity of the roles that different dynamics terms play in the dynamics equations and in 1981 and 1983, Paul extended the idea to the elimination of the insignificant dynamics terms and expressions within terms when using the equations for manipulator control. The importance of the velocity-dependent terms has been controversial and in their book of 1982, Brady et al demonstrated that there are situations where centripetal and Coriolis forces dominate the inertial forces. The manipulator joints experience high velocities during gross motions when the controller accuracy is not critical. During fine motions when the control accuracy is important, joints move with high accelerations and low velocities so that the gravitational and inertial forces become dominant and velocity-dependent forces are not so important.

As the example selected for this book is concerned with the gross motions associated with path planning and not the fine motions associated with approach paths, the inertial terms can be assumed to be less significant.

The inertial and coupling inertia terms can be excluded to give the following simplified equations:

$$\tau_1 = d\theta_1/dt\ d\theta_2/dt\ 2\{m_1(L_1/2)\cos\theta_2 - m_2L_1^2\cos\theta_2\sin\theta_2 + m_2L_1(L_2/2)\cos\theta_2\cos\theta_3\} + d\theta_1/dt\ d\theta_3/dt\ 2\{m_2(L_2/2)^2\cos\theta_2\cos\theta_3 + m_2L_1(L_2/2)\sin\theta_2\sin\theta_3\}$$

$$\tau_2 = d\theta_2/dt\ d\theta_3/dt\ 2m_2L_1(L_2/2)\cos(\theta_2+\theta_3-\pi) - (d\theta_3/dt)^2\ 2\{m_2L_1(L_2/2)\cos(\theta_2+\theta_3-\pi)\} - (d\theta_1/dt)^2\ \{m_1(L_1/2)^2\cos\theta_2\sin\theta_2 + m_2L_1^2\cos\theta_2\sin\theta_2 + m_2L_2(L_1/2)\cos\theta_2\cos\theta_3\} - m_1g(L_1/2)\cos\theta_2 - m_2gL_1\cos\theta_2 - m_2g(L_2/2)\cos(\theta_2+\theta_3)$$

$$\tau_3 = (d\theta_1/dt)^2\{m_3(L_2/2)^2\sin\theta_3 + m_3L_1(L_2/2)\sin\theta_2\sin\theta_3\} + (d\theta_2/dt)^2\{m_3L_1(L_2/2)\cos(\theta_2+\theta_3-\pi)\} - m_2g(L_2/2)\cos(\theta_2+\theta_3-\pi)$$

so that:

$$D_{12} = 2\{m_1(L_1/2)\cos\theta_2 - m_2L_1^2\cos\theta_2\sin\theta_2 + m_2L_1(L_2/2)\cos\theta_2\cos\theta_3\}$$

$$D_{13} = 2\{m_2(L_2/2)^2\cos\theta_2\cos\theta_3 + m_2L_1(L_2/2)\sin\theta_2\sin\theta_3\}$$

$$D_{1g} = 0$$

$D_{22} = 2m_2L_1(L_2/2)\cos(\theta_2+\theta_3-\pi)$

$D_{24} = 2m_2L_1(L_2/2)\cos(\theta_2+\theta_3-\pi)$

$D_{25} = \cos\theta_2\sin\theta_2 + m_2L_1^2\cos\theta_2\sin\theta_2 + m_2L_2(L_1/2)\cos\theta_2\cos\theta_3\}$

$D_{2g} = m_1g(L_1/2)\cos\theta_2 + m_2gL_1\cos\theta_2 + m_2g(L_2/2)\cos(\theta_2+\theta_3)$

$D_{33} = m_3(L_2/2)^2\sin\theta_3 + m_3L_1(L_2/2)\sin\theta_2\sin\theta_3$

$D_{34} = m_3L_1(L_2/2)\cos(\theta_2+\theta_3-\pi)$

$D_{3g} = m_2g(L_2/2)\cos(\theta_2+\theta_3-\pi)$

To determine the dynamics constants experimentally, it is important to know the joint torques for all the joints at any time instant. This can be achieved using the method described in Chapter five to monitor the joint motor currents. As the manipulator joints are assumed to be actuated by electric motors, joint motor currents provide a measurement of the torque being exerted by the joints. Figure 8.2 shows a typical relationship between a joint motor current and joint output torque.

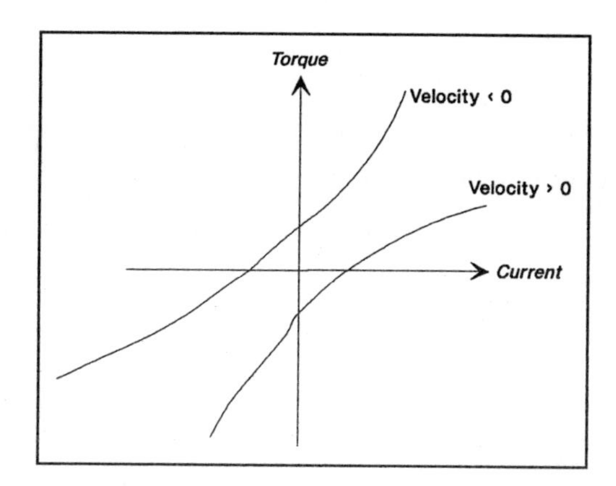

Figure 8.2: A sketch of torque versus motor current.

The output torque is approximately linear to the motor current except for an offset at the origin and a diverging curvature on both curves, which corresponded to the two directions of motion. The offset at the origin is caused by static friction that the joint must overcome before any motion at the joint can result. The diverging characteristic can be explained by the load-dependent nature of joint friction, which increases non-linearly with an increase in load. In this Chapter the functional relationship between joint torque and current is assumed to be a linear relationship so that the process of computing torque from current is a simple linear mapping and in practice the torque constants provided by the manufacturer are acceptable.

Summary: *The position and velocity were measured for various inputs. The joint torques necessary to generate motion were observed while the manipulator moved along trajectories with known motion parameters. Since the joint torque was directly related to the constants by the dynamics equations and the intermediate joint positions were known, a set of equations linear to the constants could be established from the readings of joint current and joint position and used to solve for the constants in the equations of the dynamics. This method took the non-linearity of the manipulator into account and the method is described in the next Section.*

8.4 An Experimental Method to Determine the Dynamic Model.

The procedures described in the previous Section have been applied to the base, shoulder and elbow joint of the Mitsubishi RM 501 robot with an end effector load of 2 lbs. An 80286 micro-computer controller provided torque commands to the motors through 8-bit D to A converters. The angular positions of the joints were fed back to the computer from optical encoders mounted on motor shafts. The encoder outputs were converted to a count representing position and were read by the computer via the G64 bus. Software for the system was developed in Desmet-C and then Quick-Basic. The motors were current controlled.

A series of three tests was conducted:

(i) Static tests.

(ii) Single joint motion tests.

(iii) Multiple joint motion tests.

(i) Static tests: To obtain the gravitational constants from the knowledge of joint torques, the effects due to other dynamics terms were eliminated so that the joint torque became a function of gravity loading. Only the joint of interest was moved and the other joints were stationary. Under these test conditions, the velocity and acceleration dependent terms disappeared.

With the other joints locked in a particular configuration, the torque or force required to move each joint was measured. The gravitational torques were estimated by moving the manipulator to a desired configuration and then incrementing the output through the D/A converter 1 bit at a time until motion was detected. The result of these measurements was a table of gravitational torques (D_{ig} for link i) for varying θ_1, θ_2 and θ_3.

If τ_{pi} was the torque in one direction and τ_{mi} in the other, and F_{is} represented static friction for joint i, the following equations were obtained:

$$\tau_{pi} = D_{ig} + F_{is}$$

$$\tau_{mi} = - D_{ig} + F_{is}$$

so that:

$$D_{ig} = (\tau_{pi} + \tau_{mi})/2$$

This procedure was repeated for each ten degree increment of each joint angle that occurred as a basis function for D_{ig}. Two constants, **A** and **B**, were to be determined to satisfy:

$$\mathbf{A} = m_2 g L_2/2$$

$$\mathbf{B} = g L_1 (m_2 + m_1/2)$$

so that:

$$D_{3g} = \mathbf{A} \cos(\theta_2 + \theta_3 - \pi) = -\mathbf{A} \cos(\theta_2 + \theta_3)$$

$$D_{2g} = \mathbf{B} \cos(\theta_2) - D_{3g}$$

$$D_{1g} = \mathbf{0}$$

The results obtained were unexpected and are shown in figures 8.3 to 8.12 and figures 8.13 to 8.22. The results are discussed in Section 8.6.a.

(ii) Single joint motion tests: This was achieved by driving the motors at a constant velocity. Practically, this was achieved by putting out a step velocity demand and running the joints through 10 degrees before taking any readings to avoid the inertial effects. Only one joint was moved at a time so that the governing equation was:

$$\tau_i = b_i(d\theta_i/dt) + F_i + D_{ig}$$

With gravitational compensation this could be reduced to:

$$\tau_i = b_i(d\theta_i/dt) + F_i$$

where

F_i is the Coulomb friction

b_i is the overall viscous damping coefficient.

so that the steady-state velocity was:

$$(d\theta_i/dt)_{ss} = \frac{\tau_i - F_i}{b_i}$$

The current required to maintain a constant velocity, and the velocity of the base joint for a constant demand output, were recorded for various configurations. Again the results were surprising and are discussed in Section 8.6 and shown in figures 8.23 to 8.27.

(iii) Multiple joint motion tests: To estimate the coupling terms in the dynamic equations, motions requiring joints to move simultaneously were applied. The same input was applied to joint i, first with joint j stationary and then with joint j also in motion. The response in the two cases with gravitational compensation was assumed as:

With coupling

$$\tau_{ic} = H_{ij}(d\theta_{ic}/dt)(d\theta_{jc}/dt) + b_i(d\theta_{ic}/dt) + F_i$$

Without coupling

$$\tau_i = b_i(d\theta_i/dt) + F_i$$

so that

$$H_{ij}\theta_{ic}\theta_{jc} = \tau_{ic} - \tau_i$$

where the subscript c indicates the presence of coupling.

The measured motion responses together with previously computed values of b_i and F_i were to be used to evaluate the coupling coefficients in the above equations.

In the event, this evaluation was not necessary.

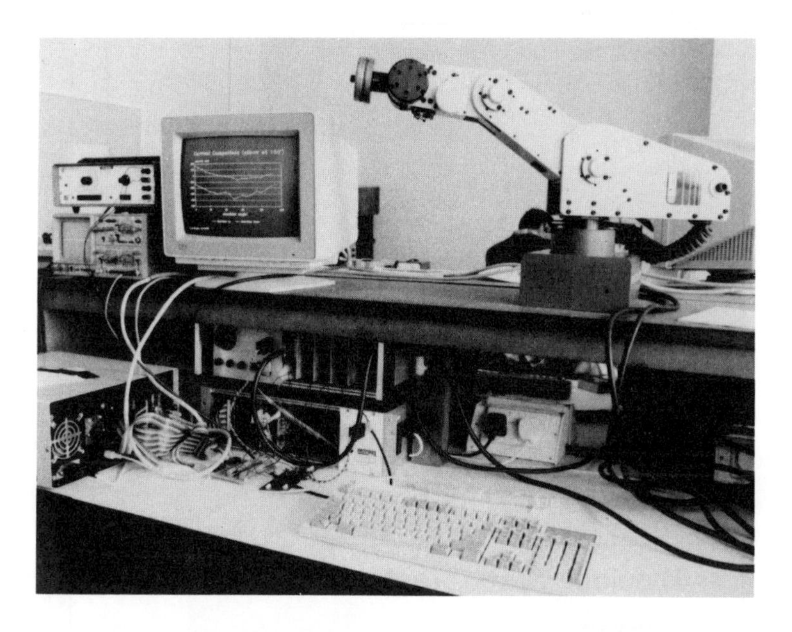

The determination of the dynamic model.

8.5 Results.

The graphical results from the static and motion tests are presented in this Section.

(i) Static Tests: The initial series of ten graphs show the shoulder current required to overcome gravity and the static friction of the shoulder joint for various configurations of the elbow joint.

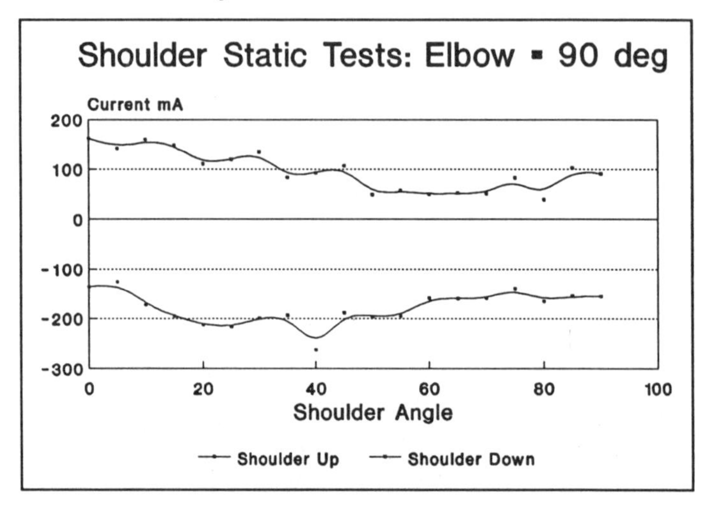

Figure 8.3: Elbow Joint at 90 degrees.

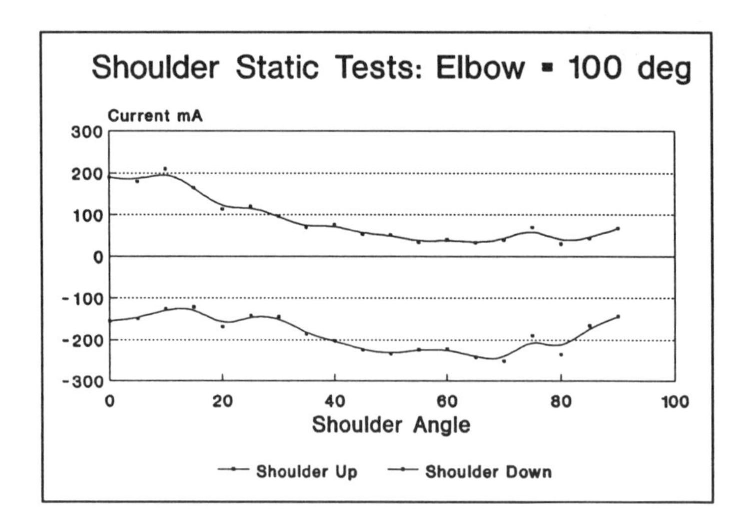

Figure 8.4: Elbow Joint at 100 degrees.

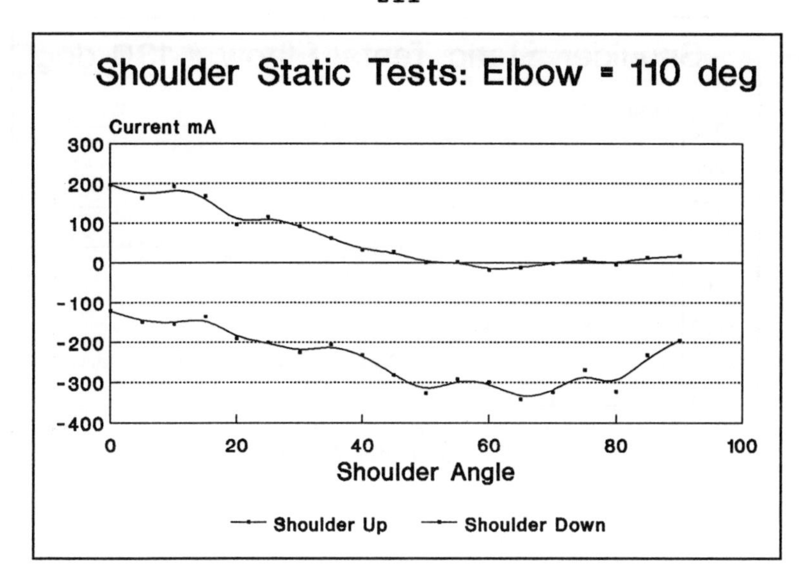

Figure 8.5: Elbow Joint at 110 degrees.

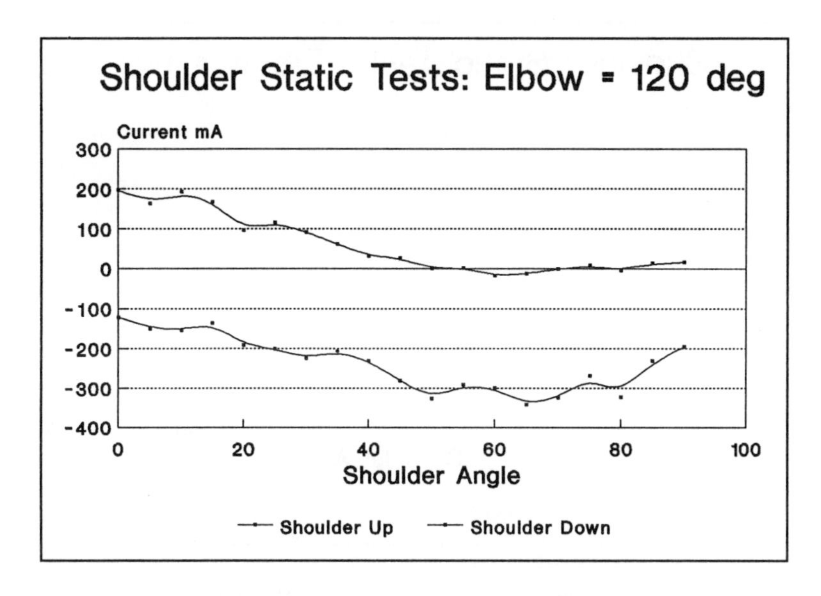

Figure 8.6: Elbow Joint at 120 degrees.

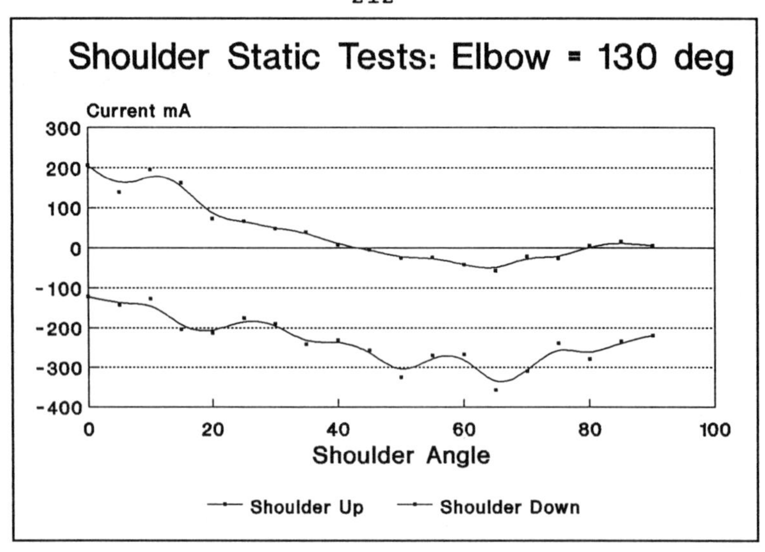

Figure 8.7: Elbow Joint at 130 degrees.

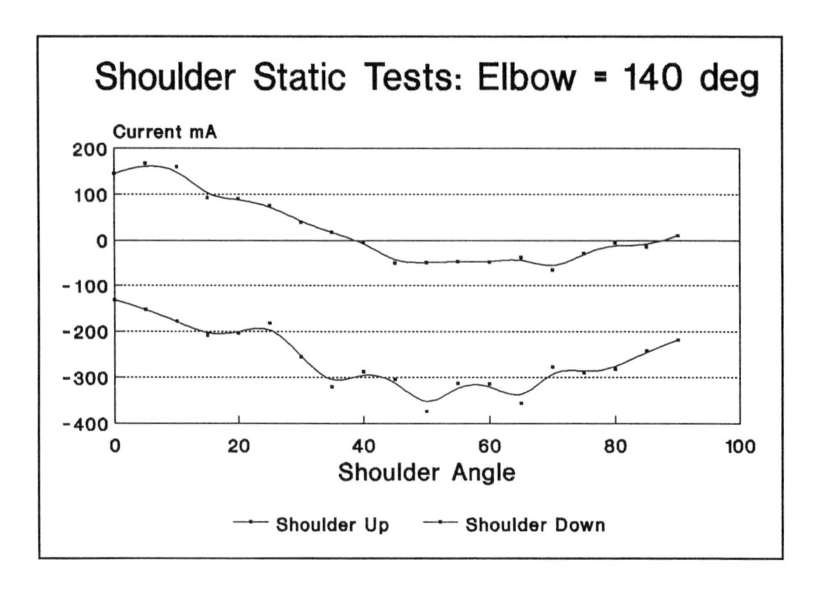

Figure 8.8: Elbow Joint at 140 degrees.

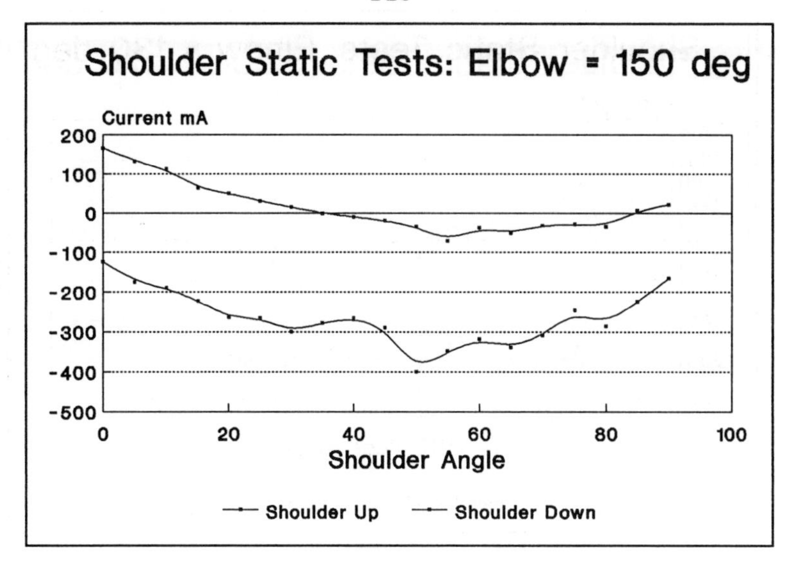

Figure **8.9**: Elbow Joint at 150 degrees.

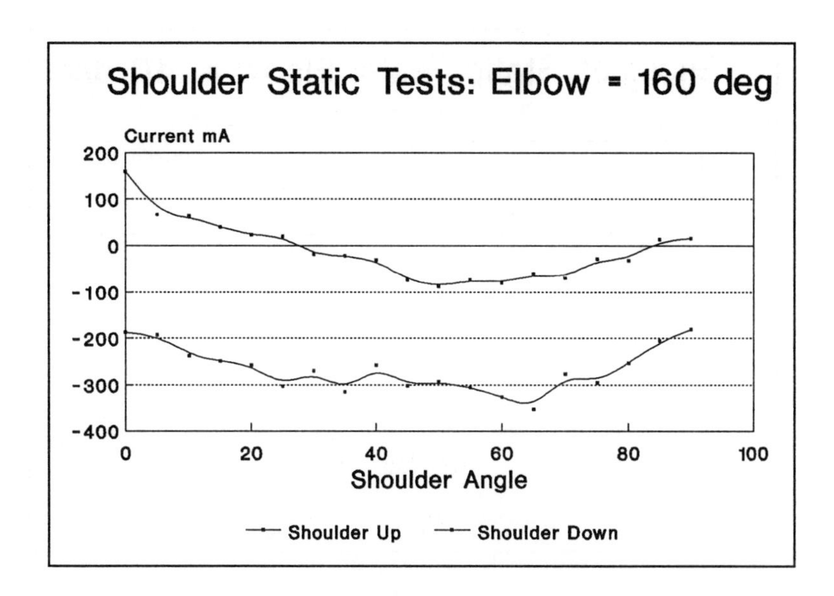

Figure **8.10**: Elbow Joint at 160 degrees.

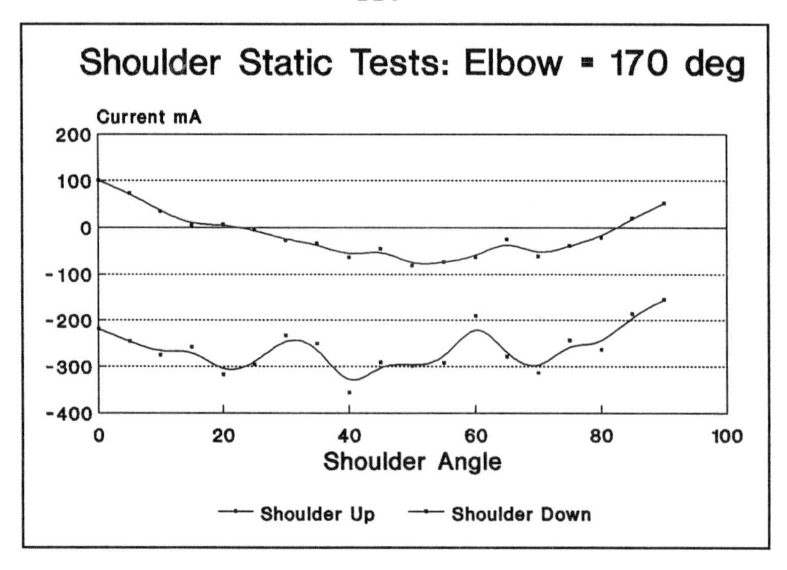

Figure 8.11: Elbow Joint at 170 degrees.

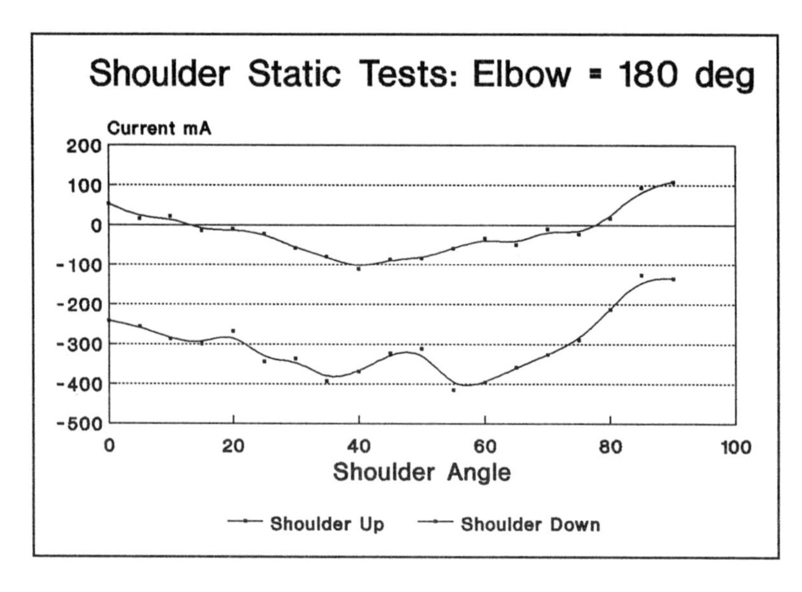

Figure 8.12: Elbow Joint at 180 degrees.

Figures **8.3** to **8.12** show τ_{pi} (the torque in one direction) and τ_{mi} (the torque in the other direction). As discussed in section 8.4, F_{is}, the static friction for joint i could be removed as:

$$\tau_{pi} = D_{ig} + F_{is}$$

and

$$\tau_{mi} = - D_{ig} + F_{is}$$

so that:

$$D_{ig} = (\tau_{pi} + \tau_{mi})/2$$

The remaining D_{2g} is shown in figures **8.13** to **8.22** with the Elbow angle marked underneath.

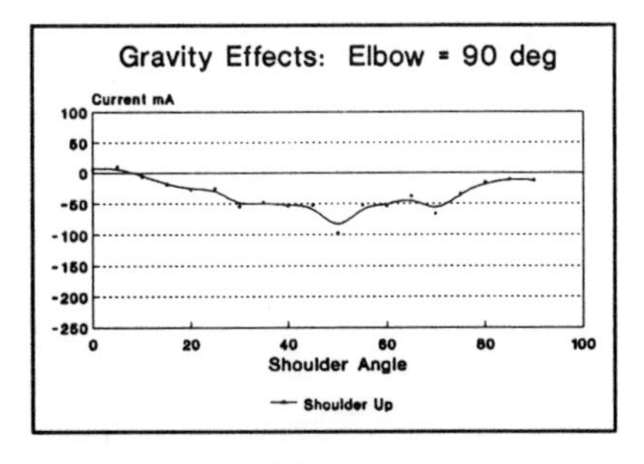

Figure **8.13**: 90

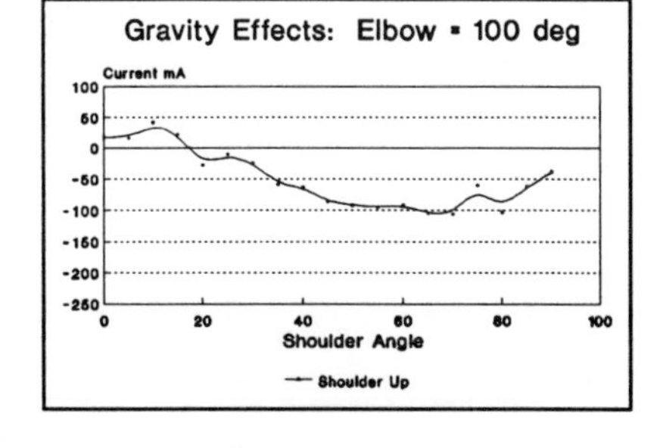

Figure **8.14**: 100

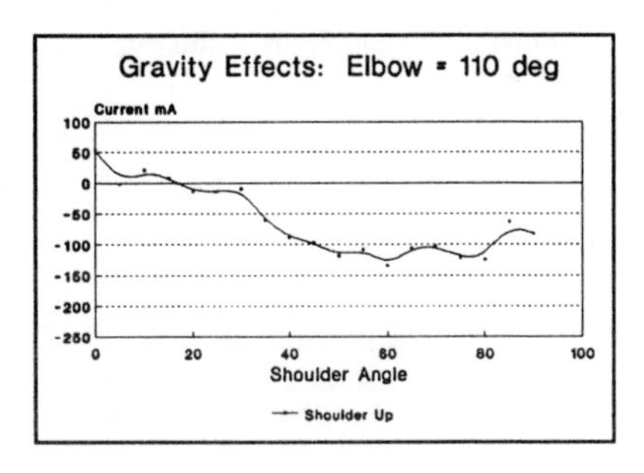

Figure **8.15**: 110

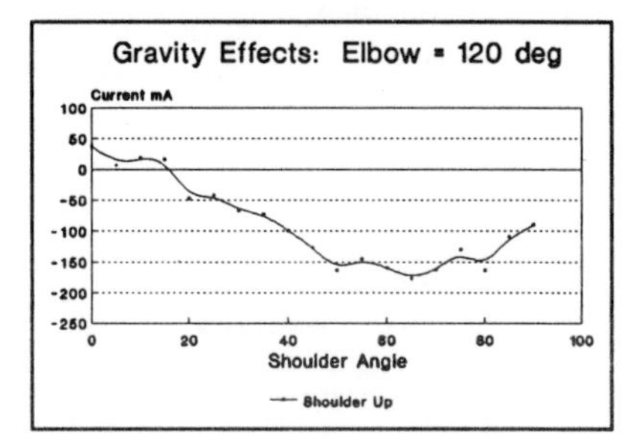

Figure **8.16**: 120

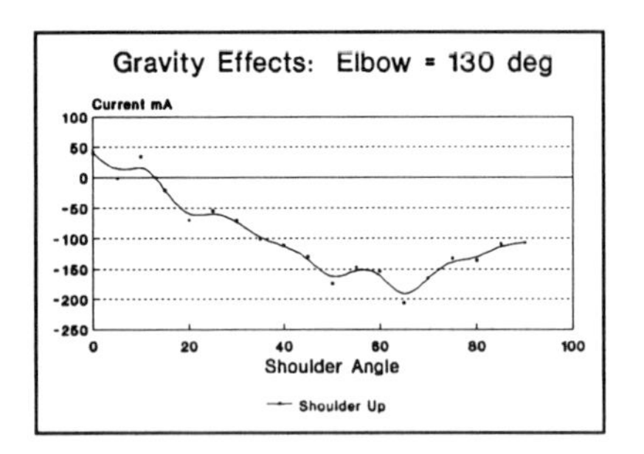

Figure 8.17: 130

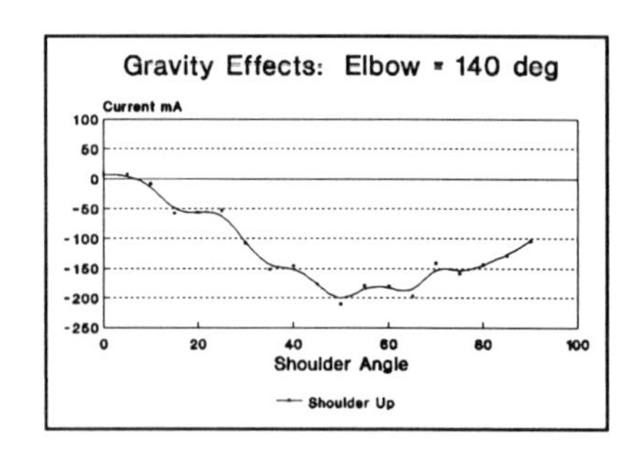

Figure 8.18: 140

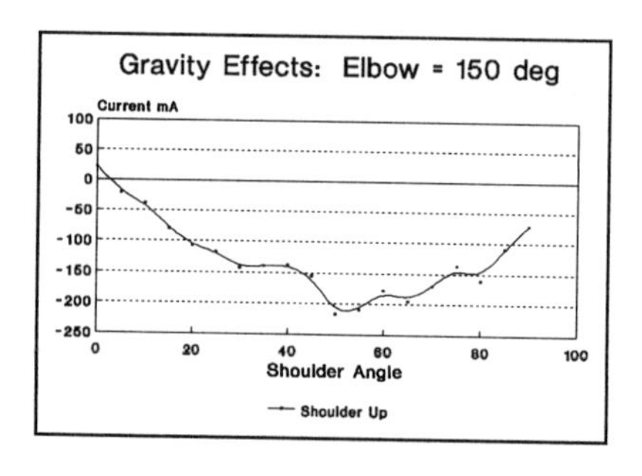

Figure 8.19: 150

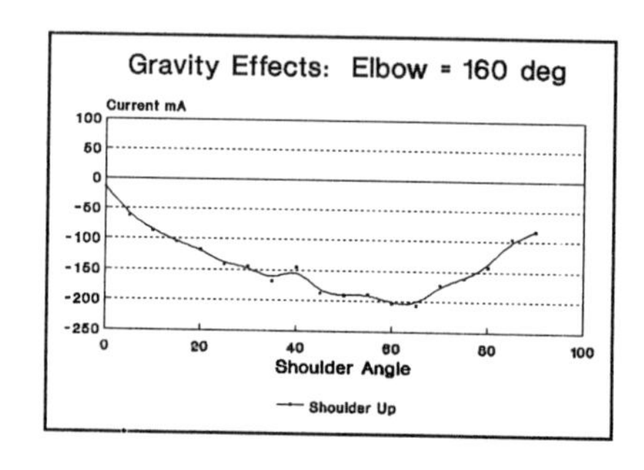

Figure 8.20: 160

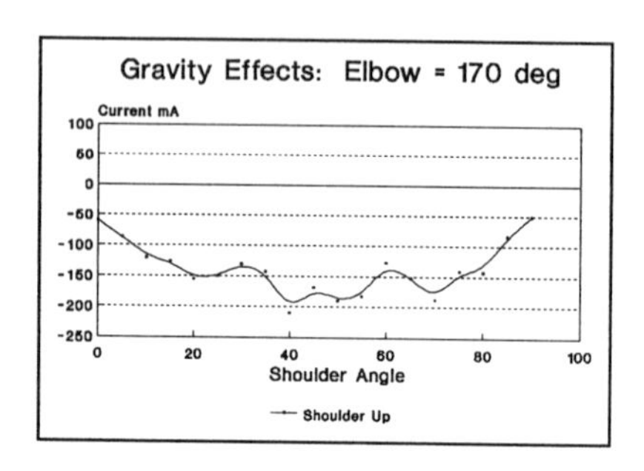

Figure 8.21: 170

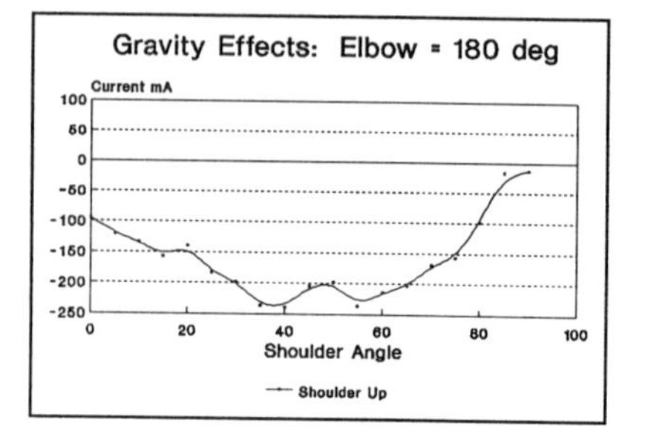

Figure 8.22: 180

(ii) Single Joint Motion Tests: Figures 8.23, 8.24 and 8.25 show the current required to maintain a constant velocity for each joint for different configurations. Figure 8.25 contained unexpected results for the base joint and this is investigated further in figures8.26 and 8.27.

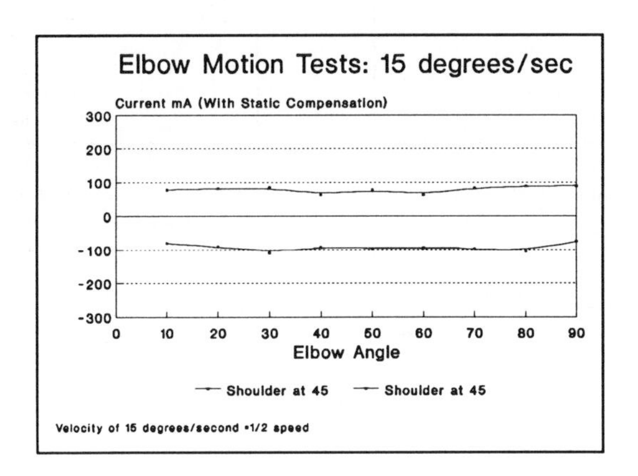

Figure 8.23: The Current required to drive the Elbow at a constant velocity.

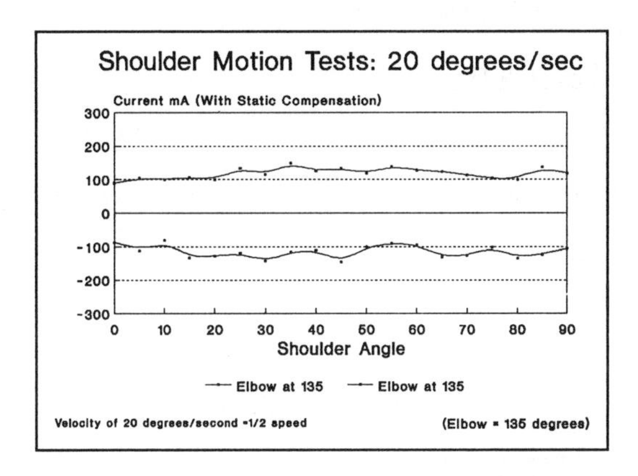

Figure 8.24: The Current required to drive the Shoulder at a constant velocity.

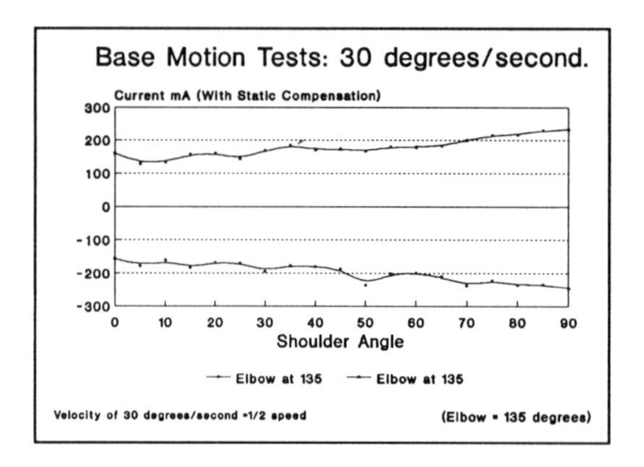

Figure 8.25: The Current required to drive the Base at a constant velocity.

Figures 8.26 and 8.27 show the Base joint velocity for different configurations of the Shoulder and Elbow.

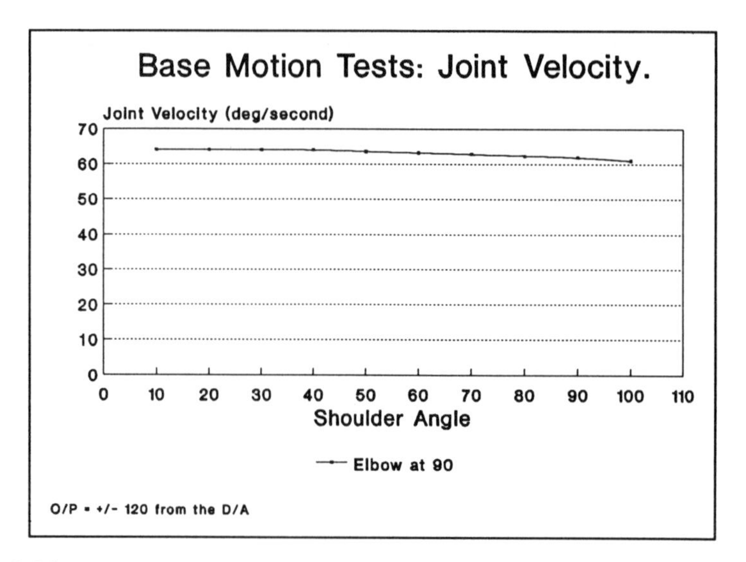

Figure 8.26: The Base Joint Angular Velocity for varying Shoulder Configurations, with the Elbow static at 90 degrees.

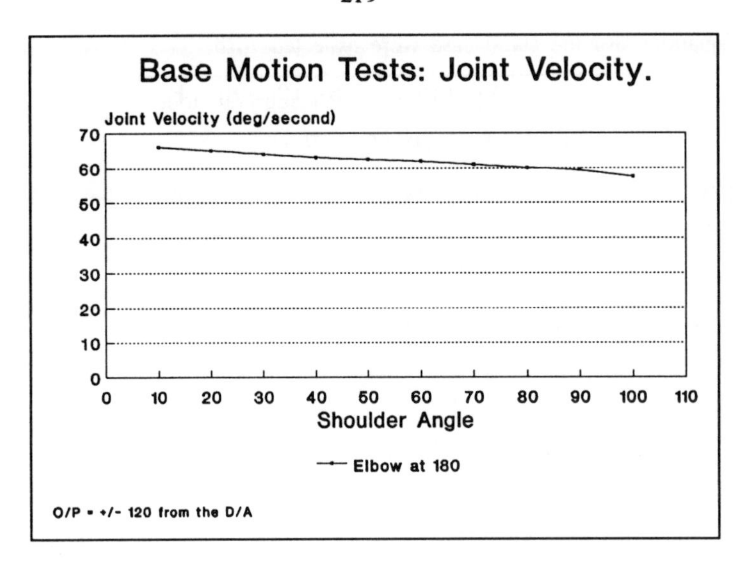

Figure 8.27: The Base Joint Angular Velocity for varying Shoulder Configurations, with the Elbow static at 180 degrees.

(iii) Multiple Joint Motion Tests: The noise in the system was greater than any effects due to coupling between joints.

8.6 Application of the Model to Improve the Motions.

This Section is in two parts:

(a) A discussion of the results.

(b) The development of Simple Rules for Path Adaption.

(a) A discussion of the results.

(i) Static Tests: The equations for the manipulator dynamics developed in section 8.4.isuggested that the maximum gravitational effect would be felt by joints θ_2 and θ_3 at $\theta_2 = 0°$. $\theta_3 = 180°$.

and the minimum effect at

$\theta_2 = 90°$. $\theta_3 = 180°$.

as the equations for the static case were expected to be

$$\tau_2 = \mathbf{B}\cos(\theta_2) + \mathbf{A}\cos(\theta_2+\theta_3) + \mathbf{F}_{is}$$

$$\tau_3 = -\mathbf{A}\cos(\theta_2+\theta_3) - \mathbf{F}_{is}$$

The practical results in figures 8.3 to 8.22 show that the maximum effect was felt by the robot at

$$\theta_2 = 40° \ \&\ 55°. \qquad \theta_3 = 180°.$$

and their were two minima, one of which was predicted at

$$\theta_2 = 90°. \qquad \theta_3 = 180°.$$

and a second at $\theta_2 = 0°$. $\theta_3 = 110°$.

Detailed inspection of the robot revealed a spring included in the robot design as gravity compensation for the arm. From inspection of the static results, the spring effects could be roughly modelled by cos of $2\theta_2$ over the range 0° to 45°, so that the equation for D_{2g} became approximately:

$$D_{2g} = \mathbf{B}\cos(\theta_2) + D_{3g} - \mathbf{C}\cos(2\theta_2) \qquad \text{for } \theta_2 < 45.°$$

$$D_{2g} = \mathbf{B}\cos(\theta_2) + D_{3g} \qquad \text{for } \theta_2 > 45.°$$

where $\mathbf{C} \approx \mathbf{B}$.

(ii) Single joint motion tests: Considering the equation from Section 8.4.ii:

$$(d\theta_i/dt)_{ss} = \frac{\tau_i - F_i}{b_i}$$

joints θ_2 and θ_3 performed as expected as shown in figure 8.23 and 8.24, in that they were not affected by the configuration of the other joints. The base joint θ_1 however, was affected by the configuration of θ_2 and θ_3. Figure 8.25 shows that the base joint had a steady state velocity which was dependent on joint angles θ_2 and especially θ_3.

It was expected that the velocity of θ_3 would have been greater as the mass moved towards the origin. The practical results show that this was not the case. In fact the opposite was true.

The inconsistency between the expected results and practical results for the base joint can be explained by considering the balancing of the robot arm and the large rear section of link L_1 which housed some of the motors. The large rear section can be seen in figure 8.28. This design meant that when the arm was extended horizontally the whole unit was balanced at the base joint, but with the arm vertical the rear Section was pulled down by gravity causing increased friction within the base gearbox. This increase in friction resulted in a decrease in steady state velocity as shown in figures 8.26 and 8.27.

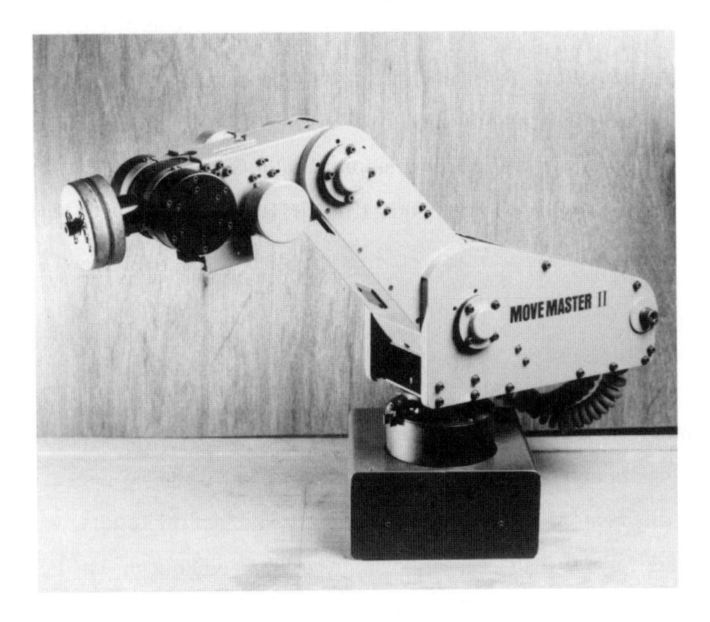

Figure 8.28: The Mitsubishi RM-501 robot (showing the large rear section housing the motors).

(iii) Multiple joint motion tests: There were no measurable velocity effects due to coupling effects between the joints. Although results were not recorded there was an obvious inertia coupling between joints θ_2 and θ_3. This could be considered in future work.

(b) The Development of Simple Rules for Path Improvement.

Considering the results of the position and velocity tests, only three effects dominated the dynamics of the Mitsubishi robot. They were:

(i) The varying effect of θ_2 and θ_3 on the friction of the base joint.

(ii) The balance spring connected to L_1.

(iii) The gravity effect of θ_3 upon θ_2.

These suggested two simple rules by which the robot path could be improved.

RULE (i) To reduce the base friction during movements of the base, the arm should attempt to balance the base mechanism by moving θ_2 towards 0° and θ_3 towards 180°.

RULE (ii) To reduce the effects of gravity loading, the arm should move θ_3 towards 90° during motions of θ_2.

Because rule (ii) has an effect on rule (i), rule (i) was given precedence over rule (ii).

8.7 Results from the Application of the Simple Rules.

Once these rules had been established, motion tests were undertaken for various paths and the times for the revised paths were recorded. The tests were repeated with three different Mitsubishi RM 501 robots and typical results were:

(i) To test for the reduction in coulomb friction: The arm was initially moved from [140°,0°,180°] to [-140°,0°,180°]via [0°,90°,180°]. The movement took an average of 4.44 seconds. When the test path was modified to use the same START and GOAL, but to move through a via-point at [0°,0°, 180°] the robot took an average of 4.14 seconds, a saving of 0.3 seconds (≈6.8%).

(ii) To test for the reduction in gravity loading: Similar tests were conducted for the shoulder and elbow, with the waist still (at 0°). The shoulder was moved from -10° to 90° with the elbow at 180°, this gave an average time of 1.94 seconds.

When the path was modified so that the elbow moved in towards 90° until the shoulder reached 50° then moved out to 180°, an average time of 1.74 seconds was recorded, a saving of 0.2 seconds (≈10%).

The adaption rules were included in the automatic path planning and adaption system and the two sets of code are shown below:

```
ShoulderDiff =Shoulder(n%+1) - Shoulder(n%)
NewShoulder(n%) =Shoulder(n%) +ShoulderDiff/2
ElbowDiff =Elbow(n%+1) - Elbow(n%)
IF SGN(ElbowDiff) =1 THEN
        NewElbow(n%) =Elbow(n%) - ShoulderDiff/6
ELSE    NewElbow(n%) =Elbow(n%+1) - ShoulderDiff/6
END IF

BaseDiff =base(n%+1) - Base(n%)
NewBase(n%) =Base(n%) +BaseDiff/2
ShoulderDiff(n%) =Shoulder(n%+1) - Shoulder(n%)
ElbowDiff =ElbowDiff(n%) - ElbowDiff(n%+1)
IF BaseDiff < 30 THEN
        IF (Shoulder(n%+1) >0) AND SGN(ShoulderDiff) =1 THEN
                NewShoulder(n%) =Shoulder(n%) - BaseDiff/2
                IF SGN(ElbowDiff) =1 THEN
                        NewElbow(n%) =Elbow(n%) +BaseDiff/4
                ELSE    NewElbow(n%) =Elbow(n%+1) +BaseDiff/4
                END IF
        ELSE IF (Shoulder(n%+1) >0) AND SGN(ShoulderDiff) =0 THEN
                NewShoulder(n%) =Shoulder(n%+1) - BaseDiff/2
                IF SGN(ElbowDiff) =1 THEN
                        NewElbow(n%) =Elbow(n%) +BaseDiff/4
                ELSE    NewElbow(n%) =Elbow(n%+1) +BaseDiff/4
                END IF
        ELSE IF (Shoulder(n%+1) <0) AND SGN(ShoulderDiff) =1 THEN
                NewShoulder(n%) =Shoulder(n%+1) +BaseDiff/2
                IF SGN(ElbowDiff) =1 THEN
                        NewElbow(n%) =Elbow(n%) +BaseDiff/4
                ELSE    NewElbow(n%) =Elbow(n%+1) +BaseDiff/4
                END IF
        ELSE IF (Shoulder(n%+1) <0) AND SGN(ShoulderDiff) =0 THEN
                NewShoulder(n%) =Shoulder(n%) +BaseDiff/2
                IF SGN(ElbowDiff) =1 THEN
                        NewElbow(n%) =Elbow(n%) +BaseDiff/4
                ELSE    NewElbow(n%) =Elbow(n%+1) +BaseDiff/4
                END IF
        END IF
END IF:         IF NewElbow(n) >180 THEN NewElbow(n) =180
IF NewShoulder(n) <-30 THEN NewShoulder(n) =-30
```

An example of initial paths and their adapted paths after applying the rules developed in 8.6 is shown below in figure **8.29**:

```
 70 ,  0 , 100        70 ,  0 , 100
150 ,-30 , 100       110 , 10 , 120
                     150 ,-30 , 100

 90 , 20 , 150        90 , 20 , 150
 30 , 75 , 125        60 , 45 , 140
                      30 , 75 , 125
```

Figure 8.29: Examples of Modified Paths.

In figure 8.29 two simple example paths are shown on the left and the result of applying the rules on page 232 are shown on the right. In both cases a via-point is generated which moves the shoulder and elbow through configurations which tend to reduce the friction on the robot base joint during motion.

8.8 Discussion andConclusions.

A novel method of path improvement has been presented in this Chapter. A method for calculating the manipulator dynamics model for a manipulator with three rotary joints based on the Lagrange formulation was presented. The model was compared to the practical results and these suggested that the theoretical model was not suitable for the case of a Mitsubishi RM-501 robot.

The model was refined through a sequence of static tests, single joint and multiple joint motion tests. The model included the effects of gear transmission and friction.

From the simplified model, two simple rules for path improvement were developed. These rules were applied to adapt the paths of Mitsubishi robots. The method reprogrammed a path during the first sequence of a set of repeated paths by adding via-points which moved the robot through more profitable configurations.

The rules developed for the Mitsubishi robots were unexpected and in the case of the rule to reduce coulomb friction was the opposite of the expected result. This is discussed further in Chapter nine.

The rules developed were specific to the Mitsubishi RM-501 robot but the new concept of using the manipulator dynamics to produce simple path reprogramming rules can be applied to any robot and many complex and programmable machines.

The results presented in Section 8.7 suggested a maximum improvement of ≈10%. In practice after considering 30 random paths, the average improvement was only 2.8%. This is a satisfactory improvement but the adaption algorithms are coarse, and the selection of the via-points could be improved in future work. When the method was used with the path planning algorithms described in Chapter six the software interfacing was clumsy and this could be improved in future work.

The software can be improved to interface more easily and quickly with the path planning algorithms and the addition of rules to include the inertias at the different joint angles would be a profitable next step.

Chapter Nine

DISCUSSION AND A LOOK TO THE FUTURE

9.1 Introduction.

Automatic programming and motion planning are broad and interdisciplinary subjects. They involve various aspects of sensing, computing, movement and manipulation. The rapid advances that have been made over the past seven years have made them exciting and there are many areas of research and discovery. This book has presented some solutions to the motion planning problem and demonstrated their real-time automatic programming and implementation. The algorithms can be expected to increase the autonomous ability of complex machines by automatically programming and reprogramming their controllers in changing circumstances and environments. The book has also explored methods of improving planned paths. Two new strategies for improvement are presented, one based on hardware monitoring of the servo amplifier currents and the second using simple rules developed from simplified dynamics equations. The underlying research work concentrated on methods of automatic path planning with constraints but during the work a novel parallel hierarchy control system evolved. The original concepts presented in this book include the following:

The use of diverse models for different parts of the work place. The models of the static environment were complex but accurate, while the dynamic obstacles were modelled in a fast and simple way.

The use of simplified models of the robot dynamics to improve a robot path. Until now the engineering research work described in the literature has tended to use the dynamics at lower levels in the control hierarchy to adapt a robot trajectory. The work described in this book has crossed the artificial boundary between computer science research and engineering research.

The use of monitored actuator torques to adapt a robot path. Although joint motor currents have been investigated in the literature, no attempt has ever been made in past work to use this information for path improvement.

The use of 2-D slices to enhance the speed of modelling obstacles in a joint configuration space. The use of this simple and novel modelling method increased the processing speed of the path planner.

9.2 Discussion and Review

(a) **Modelling of a robot and obstacles.**

The static environment: The static environment was modelled accurately as several polyhedra for use with the Global Planning system. This model was transformed into joint space before planning with dynamic obstacles. As this transformation took place once, at the beginning of the program, there was little or no time constraint.

The robot geometric model: The robot geometric model consisted of two lines connected at the elbow joint surrounded by a skin a constant distance from this skeleton. This model was simple and proved to be fast. The use of different models for different parts of the work place is one of the novel concepts presented in this book.

Dynamic obstacles: Several different models were considered for the dynamic obstacles and two were selected as they performed the transformation into joint space in the fastest times. The two models are spheres and 2-D slices in joint configuration space.

(i) **Spheres**: The local path planner performed faster when using spheres compared with the other models. It should be noted that there must be a point at which increasing the number of spheres, in order to increase the accuracy of a model, becomes impractical; and at this point the Data Processor could change to use a polyhedral model.

(ii) **2-D slices**: Although spheres provided the fastest performance for the local path planning algorithms, 2-D slices proved to be faster to transform for the global path planner. This was due to a large amount of the complex processing being replaced by a simpler copying function.

Other models: Of the other models considered, none performed favourably with the local path planner, but the parallelepiped and the sphere provided favourable results with the global path planner.

(b) **Image data processing and the vision system.**

No claims for novelty are made for the vision system, and many improvements could be made to this component. Some of these improvements are discussed.

The configuration of the apparatus: The configuration selected was a single camera placed above the work area. The camera placed at an angle, and the use of two cameras, were rejected because of the complexity of the processing required. Placing the camera directly above the work place allowed a simple mapping in the X-Y plane, and the later use of templates in real time required this simpler processing.

Initially the light source was placed behind the camera but later work used back lighting below the work place. This part of the system could be improved and for future work it is the intention to use a series of pictures for processing rather than single discrete pictures. This may allow the use of a light source above the work cell and the introduction of stereo vision techniques.

Low level vision techniques: The low level processing was performed within the bounds of a small window (3 pixels x 3 pixels), and had no knowledge of intensities outside this window. All the methods performed satisfactorily, but gray level weighting was excluded because of the time taken in processing. It will be practicable to include this aspect in the future, using faster, dedicated processors.

High level vision techniques: The higher level techniques aimed to interpret the data supplied in the form of edges and regions of some known object. This relied on some concept of 'intelligent' processing, that is, the ability to extract pertinent information from a background of irrelevant detail. The edge detection method used was one of the simplest forms of "intelligent" processing in that it extracted pertinent information regarding the position and connection of edge points.

(c) **The systems and the apparatus**

The apparatus was developed over a period of five years. During that time the state-of-the-art of computer hardware has advanced, and future developments of the work described in this book would benefit from the use of transputer arrays of parallel processors.

The systems: The sub-systems worked together satisfactorily. The systems were designed to work in parallel in different computers and could move easily to a new parallel apparatus.

The communications sub-system: Investigation of the communication between sub-systems revealed that communications speed was not a significant limiting factor compared with the time taken for the complex processing in each computer. For this reason, and to use the interrupt facility, the two standard RS 232 ports available on each micro-computer were used. For future work, communications may be simpler, as modern parallel processors and parallel programming languages are designed for fast communication between processors and processes.

The G64 bus: The G64 bus was adequate for the work described in this book but proved to be a limiting factor as clock frequencies through the bus were limited to 1 Mhz. It is the intention to expand the system to program three robots and to control other machinery. The G64 bus is not adequate for this purpose and is being replaced.

The Mitsubishi RM.501 robot: The robot proved to be an interesting choice as the dynamics were unexpected and surprising. The robot had a limited reach and work area, and it is the intention to expand the system to use a Unimation Puma robot and either a Fanuc 600 series robot or Syke 600-5.

The robot controller and servo-amplifiers: The controller and servo-amplifiers worked satisfactorily and they are now being redesigned for use with other robots.

(d) **Automatic path planning.**

Planning in 2-D space: Two methods were developed on an initial test rig (a prototype base), one a local and heuristic method and the other a global method. These methods worked satisfactorily in 2-D SPACE when considering a simulated second joint and link, but it was not until the methods were extended to 3-D SPACE using the Mitsubishi robot that physical results could confirm the expected results.

Local heuristic methods in 3-D space: The local heuristic method worked within the definition of real time used in this book but generally the planning took twice as long as the global method. As the advantage of a local planning method over a global planning method should have been a faster speed of operation, and this was not achieved, it is not intended to extend this method in the future.

The path planning process took less than 3 seconds, and a large proportion of the planning time tended to be taken up in considering the static environment. Any future work could consider methods of speeding up this part of the process. The method would be useful if a global model of the work area were not available, for example in undersea or space applications.

Global Method in 3-D space: The global method used a discrete range of values for each degree of freedom, namely five degrees. If the range were extended to ten degrees, then the number of units would be reduced by a factor of eight and the calculation time could be reduced by a similar factor. The path planning process took less than 0.5 seconds.

For simplicity, the range of values for each degree of freedom was set to the same value. In practice particular degrees of freedom may be more important than others. Smaller ranges of values could be used for the more important robot axes, for example, the base angle Θ_1 in the case of a revolute robot.

In the future, when manoeuvring a work-piece close to obstacles, the degrees of freedom of the gripper (in this case Θ_4 and Θ_5) could be considered. This would create a graph of more than three dimensions. The disadvantage would be the size of graph, but it is intended to extend this work to six degrees of freedom for other robots and for use in real-time-flight-simulation systems, and for advanced agricultural systems.

A dynamic size of graph could be used. In large areas of either CLEAR or BLOCKED nodes the unit ranges could be larger, but in the areas around the surfaces of obstacles the graph could use smaller units. The processing to achieve a dynamic graph may be complex, but it is the intention to experiment with dynamic graphs on the new apparatus.

The performance of the system was encouraging in that the robot could calculate and recalculate paths quickly (<6 seconds after introducing an obstacle into the work place). Performance for the path planning methods was difficult to quantify as no other working systems existed to use as a Bench Mark. In 1986 Khatib presented work which described a collision avoidance system which worked in near real time, and video film of the system working with a robot is held by the BBC. The work presented in this book compared favourably with the system shown on the video.

Trajectory generation: The trajectories generated were relatively simple and it is the intention to use more complicated cubic (or higher order) splines for future work.

(e) **Sensor fusion.**

Sensor fusion concerns the integration of data from two or more sensors. Conventionally, sensors are classified into internal state sensors and external sensors. This book has dealt mainly with external sensors. The data from these sensors was used to create a unified representation of the sensed environment. A prototype sensor system was developed and different sensors were integrated.

Conflicts were detected when more than one external sensor was used and the major conflicts tended to be due to differing material types or differing ranges. The conflicts fell into two broad divisions: the same technology providing conflicting information; different technologies providing conflicting information. This was dependent on the type and number of sensors in the system. Obstacles detected by the system were assigned certainty levels and placed on a grid which described the 3-D environment.

The choice of which sensor to use is often based on experience or on considerations such as cost or availability. More rational choice requires modelling techniques that use a better description of a sensor's abilities and limitations. This book has described methods of combining uncertain sensor measurements to provide useful information in the event of contradictions. For the future, parallel architectures may be required for real-time sensor control, as sensors are often distributed in space and among different time frames.

Types of sensor fusion: There are three types of sensor fusion: competitive, complementary and independent, and of these, complementary and competitive fusion types were considered. The efficient integration of data from two or more sensors into a unified representation of a sensed environment has been discussed. Different sensors were integrated and some initial results presented. Obstacles detected by the system were assigned certainty levels and placed on a grid which described the 3-D environment.

The test rig to investigate sensor fusion: The test rig used three different technologies: structured lighting using a laser light source, an ultrasonic proximity switch and limit switches. A structured lighting system was created, which consisted of a camera interfaced to a micro-computer and a laser light source. Gray levels of the image were reduced to eight levels before being stored. The image was smoothed, a threshold was applied, and the edges of the laser spot were detected. The system provided range information from 85cm to 175cm. Median filtering was used to reduce noise and other spurious effects.

Fusing the sensor information: The fusion algorithm was based on the interaction between the source corroboration and the principle of belief enhancement/withdrawal. The method was tested and certainty grids were used to represent the obstacles by dividing the environment into elements (denoted as cells). Each cell contained a certainty value that indicated the measure of confidence that an obstacle existed within a cell area. This method provided a reliable obstacle representation in spite of sensor inaccuracies.

The test programs combined six different technologies. Simulated technologies included an inductive switch which detected an object within the range 0-15mm. A capacitive proximity switch detected steel within 70mm of the sensor, cardboard within 28mm and glass within 35mm of the sensor. An optical proximity switch detected steel within 520mm and cardboard within 120mm of the sensor. The test rig included limit switches which detected when the obstacle range was zero. The laser system sensed cardboard or steel within the range of 400-1500mm. The ultrasonic system sensed steel or glass in the range 400-3000mm.

The program considered the maximum and minimum sensing range for different materials and colours, and a percentage reliability was generated for each sensor. If the reliability of the sensor was deemed to be low then the output from the sensor was ignored. The information from the sensors was categorized into: touching, near and far. This categorization established the necessity for corrective action.

Problems occurred as conflicting information was received from the sensor systems. These errors were classified by extending the certainty grid to include variable flags for each cell which indicate when conflicts occur. The cells where conflicts were identified were set to BLOCKED. In future systems the programs can be modified to learn from these conflicts. A database was produced containing the learned conflicts and this was referred to when conflicting information was received. Known conflicts were effectively resolved.

Typical conflicts: Conflicts tend to occur when: an obstacle is out of range of one sensor and in range of another, obstacles are made of materials that cannot be detected by some sensors, or because of directivity errors. Depending on the sensors selected, different types of conflict can occur: same technologies giving different opinions; different technologies giving different opinions. A list of typical materials which can provide conflicting information has been presented. Another kind of sensor conflict occurred when a sensor detected a non-existent obstacle or when operational errors occurred.

The Laser system was limited to a controlled environment. To overcome this, the vision software may be made more complex, but there is still a limit to the improvement that can be made. Methods of obtaining range information from the camera and the laser can be refined to increase the speed of operation. One simple change is to reposition the laser above the camera instead of alongside.

Range recovery through camera motion is being investigated in some research laboratories; this does not require additional hardware, but it does require additional complex software.

Programs are able to refer to past conflicts by interrogating a database. A simple decision graph and a set of simple rules resolve these remembered conflicts. Noise in the system can be handled by certainty grids but one set of conflicts cannot be resolved: those caused by sensor failure. These failures can be simulated and the decision graph can assign remembered results from the database to the conflict. In the further investigation of these conflicts, learning algorithms will need to be extended to deal with the failures.

(f) **Force sensing.**

Chapter five covers methods of force sensing, and methods of adapting motions to produce faster and more efficient robot trajectories have been explored. The overseer described in chapter three was modified to receive information from a Peak Detector which monitored the motor drive currents.

Monitoring of the motor drive currents: Motor drive currents were monitored to give an indication of the actuator torques. Monitoring of motor drive currents is not in itself novel but the ways in which the information is used are original. Future work could consider other methods of measuring the joint torques.

Motion improvement using force information: Current transients were detected by considering the level and gradient of consecutive current samples. Once the data from the actuator currents had been analyzed, during the next repetition of the set of movements the joint trajectories were adapted by changing the controller look-up tables. Paths can thus be modified to remove some current peaks in the motor circuits.

The software: Various non-linear control algorithms in sub-processes were loaded into the controller. Included was the optimum solution developed for the system by non-linearising the experimentally achieved, critically damped, control algorithm. These were used to produce look-up tables in memory. A repeated sequence of robot moves was entered by a human operator. During the first sequence the optimum solution was used for actuator control. As each joint angle target was reached and passed, the controller signalled the main computer. The main computer sampled the D.C. current from the servo-amplifier driving the actuator, and this information was passed to the path adaptor level. The path adapter advised the overseer of possible changes to the joint trajectories and this information was passed to the supervisory level in the controller.

Obstacle detection using force sensing: Tactile sensors that allow the computation of shear, stress and strain have been slow to emerge. This book explores methods of obtaining this information. Transients in the current waveforms are experienced when a motor is overloaded or when a link meets an obstruction and is forced to stop or slow down. The forces exerted in Cartesian space can be related to forces in the joint variables by a Jacobian matrix and these joint forces used to detect collisions by monitoring the joint motor currents. Once a machine has made contact with an obstacle, the surface of the obstacle can be followed by monitoring the motor drive currents and including them in a feedback path within the controller.

Limited collision detection is included. The system described is capable of modification, so that in the event of a collision, it would retrace the collision path and return to the previous set of trajectories by restoring the look-up table selection.

In future work, some active compliance might also be achieved by considering the joint position errors and the joint forces. This active compliance may not be sufficient for difficult assembly tasks, but could aid specialized remote centre compliance devices or robots such as the IBM Selective Compliance Assembly Robot Arm (SCARA).

(g) **Path improvement to minimise peaks in joint motor currents.**

The motor drive currents were monitored in an effort to improve the machinery motions.

Adaption to reduce changes in joint direction: The system successfully improved some robot paths and the method can be expected to extend the working life and service intervals of the servo motors and machinery, and in some cases increase the speed of operation.

In 1984 Scott reported that robot maintenance can be up to 10% of the original purchase price every year, and any reduction in maintenance costs or down-time can have a substantial effect on the investment return or payback period. Minimising the current and torque transients reduces some mechanical forces and stresses in the system. This should increase the up-time by extending the mean time between failures, and maintenance may be required less frequently. Minimising current transients results in energy conservation, allowing robots to run for longer periods from a given power source. This may be an important concept in future mobile robots or for complex machinery in inaccessible environments. In 1986 Craig reported cases when this was important, as time wasted recharging may be uneconomical and power pack replacement may be impractical.

The method of improvement was never successfully interfaced with a path planning system which included dynamic obstacles, and it would not appear worthwhile to pursue this method further.

(h) **Path improvement considering the robot dynamic equations.**

Development of a model of the robot dynamics: It was necessary to produce specific rules for the unusual design of robot selected. This was achieved by carrying out tests on the robot to calculate the torques required to move the robot at various velocities and positions. The rules developed were specific but the methods and concepts can be applied to any manipulator.

Minimum time paths are in general similar to minimum distance paths, but the shortest, most direct or most obvious path may not be the quickest. The minimum time path can be expected to take unusual routes and the greater the degrees of freedom of a robot, the worse may be the link coordination from ad hoc motion planning. Conversely the greater the number of degrees of freedom, the more possibilities there are for adaption, and the greater the improvement possible.

Adaption using the dynamic model: The rules developed for the Mitsubishi robot were simple but had some effect, with, on average, a 2.8% improvement. The robot consisted of links which could be made to work together if kinetic energy and momentum were not wasted. The links exert reaction forces on one another that are generally harmful, but it may be possible to plan paths to minimise these effects, perhaps so that links can give helpful kicks to each other at the right times; {parametric resonance}. Future work may consider the inertia parameters for the Fanuc 600 series robot as a first step towards this improvement.

A robot provided with a degree of autonomy gave result similar to human workers adapting a repetitive task to make movements easier and less tiring. The methods crossed the boundary between engineering and computer science research in that the manipulator dynamics were used at a level higher than that usually considered in engineering research. Computer science research has tended not to consider the dynamics of moving objects in the planning algorithms.

9.3. A Look to the Future.

In an effort to automate a manufacturing process, engineers have often copied the manual process. The manipulations of human arms and hands have been imitated by mechanical means. Recent research suggests that direct copying may be less successful than other, sometimes simpler solutions. The explanation may lie in the facts that the human arm and hand tend to work with many other senses and systems, and that the number of degrees of freedom achieved by the human hand has never been achieved by a mechanism.

This book has been concerned with the integration of several sensors and systems; crossing the divide that has existed between computer programmers investigating methods of automatic programming and engineers investigating machinery motions. While the work described in this book was in process, the computing power available to programmers and engineers has been increasing at a rapid rate.

Considerable research has been reported into the design of real-time systems and the development of machinery programming languages, yet many problems still remain. Trends in computer science towards object-orientated languages and parallel processing can be expected to change the way in which machinery is programmed. At Portsmouth, transputers and methods of parallel processing are already changing the way in which programs are developed and written. Objects are being described as data structures called classes, and permissible operations on the data structures are described within procedures. This allows inheritance as new classes can contain the procedures of previous classes. The result can be a considerable increase in the productivity of the programmers. Object-orientated code is more flexible than hard code but may be less efficient. For the future it is flexibility of code that will be important.

Occam processes can communicate using message passing. When one process finishes a calculation the results can be sent to the next process in a pipeline. These processes can be synchronized using rendezvous techniques as reported by McKerrow in his book "Introduction to Robotics".

Complex machines can be programmed to do a variety of tasks. From the point of view of artificial intelligence, robotics is the intelligent connection of perception to action. Within a perception model, five processes can be identified: measurement, modelling, perception, planning and action. The key to the future of motion planning and automatic programming is held within the improvement to all of these processes.

In a similar way, complex machinery can be decomposed into a set of functional sub-systems, but it is still the software that binds these sub-systems. If advances are to be made, then research must cross the traditional boundaries between these areas: mechanical, electrical, electronic, production, planning, control and computing. Only then can the necessary advances in motion, articulation and software compliance be made.

If machines are to become autonomous, problems in the areas of sensing and navigation must be solved. Dead reckoning calculated from odometry may be desirable with complex sensor systems, but as discussed in this book, as sensor systems become more complicated, the information presented from the sensor fusion becomes more difficult to interpret. Just as humans make real mistakes, we may have to accept that machines must make real mistakes if they are to learn. Until now, accurate measurement has been fundamental to the successful application of robots and complex machines but in the future measurements may be less accurate, but the data may be handled by more complex algorithms.

In future environments control systems must execute planned sequences of motions and forces in the presence of unforeseen errors. Suitable controllers may be non-linear cancelling, adaptive, and model-based controllers. The parallel processors discussed earlier should allow the solution of inverse dynamic models in real time and inverse mapping for feature-based visual servoing systems. Strategies for force control include stiffness, damping, impedance, explicit force and add-on force control systems which modify position references. Hybrid control addresses the problem of mapping force and position control in independent degrees of freedom.

New architectures will be required for the real-time control of autonomous grippers, hands and the complex machines of the future. A number of multiprocessor architectures have been designed and implemented but a different approach may be required. Work at Portsmouth by Strickland, Tewkesbury and Hollis has suggested that transputer arrays may be suitable.

The relationship between artificial intelligence and robotics has been studied by Brady in his book of 1989; both disciplines provide techniques important to automatic programming and motion planning for complex machinery. Search processes are being improved with work concerning symbolic reasoning and uncertainty. Recent work by Billingsley, Haynes and Sanders at Portsmouth suggests that neural networks may provide advances in these areas.

The greatest advances may be made in the areas of reasoning and real world planning. Early AI work concentrated on off-line planning methods in an ideal and simple world. A number of important issues were identified in these artificial worlds, including the need for hierarchical planning. The issues now being recognised as important include: the temporal planning of time critical events, planning in uncertainty and real time motion planning. Each of these requires some considerable research effort to provide workable and generic solutions.

Recent work in qualitative reasoning is suggesting that geometric, kinematic and other physical models may not be adequate for more complicated tasks requiring reasoning and intuition. Instead, more qualitative and symbolic representations may be appropriate. Brady quotes the case of reasoning about whether or not a rusty screwdriver with a chipped blade might be adequate for removing a partly worn screw. Most robotics reasoning systems ignore shape, relying on geometric models such as simple polyhedra, yet shape and symbolic representation appear to be central to our ability to make sense of the world. For example, "if you remove a particular tin of baked beans then the pyramid of tins on display will collapse".

Production engineers complain about the changes in design required to transfer a prototype from development to manufacture. Optimising the manufacturing process often involves interfering with the design of an article. Future design systems can be expected to take account of the production and manufacturing processes. One such project is "design by manufacturing simulation" at the University of Portsmouth. This work was begun by Professor John Billingsley and is now being continued by Jaques and Harrison. In this research, products are designed by a series of simulated operations on the raw materials. This method ensures that products can be manufactured with the available machinery in an efficient and effective manner. The output from the system is a set of actions, an operation sequence and the conditions under which the manufacture must take place. Automatic programming and motion planning techniques can then be used to help in manufacturing the articles, but this step may require advances in geometric reasoning.

Most of the early research into geometric reasoning has concerned robot navigation or assembly. More recently Taylor and Brooks have been working with uncertainty and this research is demonstrating that we still cannot automatically plan realistic assembly operations.

Creating an autonomous system to perform challenging tasks is complicated by the many separate aspects of the problem. Future systems need to work in real time and must degrade or upgrade gracefully as alterations and/or additions are made to the system. A further sub-problem may be the communications required between distributed processes or processors.

This book has attempted to address some of the main issues and problems requiring solutions if this dream is to become a reality, which of course it will.

Appendix A

The Transformation Programs

This Appendix describes the two programs which transform the dynamic obstacle models into joint space from cartesian space.

Setting of the lists: The first set of angles returned during the programs TransformSphere.BAS and TransformSlice.BAS were for robot collision with the centre of the sphere. The ForeFill flag was set for the " Expandout" routines. The array NodeStatus was a status register which set flags to give the status of a set of each node. This was set to BLOCKED (*ie* bit 2 was set to 1). The angles were stored in an array called List1 as shown below. The upper arm was tested to see if it would collide with the sphere in any configuration and the nearest and furthest points of the sphere were calculated.

```
NodeStatus(t1%, t2%, t3%) =2
CALL PutonList(t1%, t2%, t3%)
NearestDistance%    =L3% - Radius%
FurthestDistance%   =L3% - Radius%
```

The upperarm was tested against the sphere model to see if it would collide with the furthest point on the sphere. If it collided then the forearm was not tested and the ForeFill flag was set to FALSE. The node was removed from the list as shown below.

```
IF FurthestDistance%   <UpperLength%  THEN
     ForeFill% =false%
     CALL GetoffList(t1%, t2%, t3%)
END
```

The upperarm and sphere model were tested to see if a collision occurred with the nearest point of the sphere. If a collision occurred, the ForeFill% flag was set and the upperarm was set to point at the sphere's centre (Sphϴ). Θ_3 was set to 180° and the angles were loaded onto the list. As the upper-arm collided with the sphere, all possible Θ_3 angles would also collide. Θ_3 was set to BLOCKED between its limits for the specified Θ_1, Θ_2

```
IF NearestDistance%   <UpperLength%  THEN
  UpperFill%  =true%
  t2% =Sphϴ%
  CALL PutonList(t1%, t2%, t3%)
  FOR Loop1% =LowLimit(t3%) TO HighLimit(t3%)
    NodeStatus(t1%, t2%, Loop1%) =2          ; Set to Collision
  NEXT Loop1%
```

The flag register NodeStatus was tested to see if the particular node had already been tested by consulting bit 4 for the forearm test and bit 8 for the upperarm test.

```
IF (NodeStatus(t1, t2, t3) AND 4) =4 THEN Foretested  =true
IF (NodeStatus(t1, t2, t3) AND 8) =8 THEN Upptested  =true
```

If the flags were not set and the NodeStatus was not set for an old obstacle, then the upperarm end point cartesian coordinates were calculated using the formula's for the forward kinematics solution described in section 3.7. If the forearm was to be tested then the Foretip position in cartesian coordinates was calculated and the NodeStatus flag was set to forearm tested.

The distance between the centre of the sphere and the end tip of robot was found and a test was conducted to see if the distance was less than the sphere radius plus the sphere model for the robot. If true, the node was placed onto list1 and set to BLOCKED. The same test took place for the upper-arm. If a collision occurred with the upper-arm then the procedure was repeated.

The subroutine Expandout tested all the nodes around the reference node using the subroutine TestPos. An example is shown below for the waist joint. The joint is set to -5°, +5° and then returned to the reference node.

```
E1% =E1% - 1                  ; setting to -5° of the ref node
IF E1% >=Low Limit THEN CALL Testpos
E1% =E1% +2                   ; +10°now it's +5°to the ref node
IF E1% <=High Limit THEN CALL Testpos
E1% =E1% - 1                  ; Resetting back to ref node
E2% =E2% - 1                  ; As before except now its the
                              ; upperarm and the process repeated
```

The forearm (E3%) was only tested for the forearm fill in by testing the flag TestType, which was passed from the subroutine FillIn. The subroutine expanded each node where a collision had occurred. Before this expansion the angles were removed from the list so that it was not expanded again. This was repeated until no further collisions occurred.

The first part of the subroutine checked whether there were two nodes on the list. If there were, the last one on the list would be the upper-arm node and this was transferred to a temporary array called list2. The nodes left on the list were the forearm nodes. The flag BothArm% was set to true so that after the forearm expansion the upper-arm node could be transferred back to list1.

The ForeFill flag was tested. If it was set to true then the FillIn for the Forearm was activated. The testtype flag was set to Foretest so that when calculating the forward kinematics in the subroutine TestPos, the routine knew that the Fortip needed to be calculated.

The node was removed from list1 and passed to the subroutine ExpandOut where the node was expanded and added to list1 if it collided with the obstacle. The routine continually removed nodes from list1, expanded them and tested for collisions, until no more collisions had occurred. (list1 became empty). This routine is shown below.

```
IF ForeFill% =true% THEN
 TestType% =ForeTest%
 DO
  FOR Loop1% =1 TO NoNodesList1
   CALL GetoffList(t1%, t2%, t3%)
        CALL Expandout
  NEXT Loop1%
 LOOP UNTIL NoNodesList1 =0
```

This routine was also used for the upper-arm with the flag TestType set to UpperTest.

The limits of θ_1, θ_2 and θ_3 at which collisions occurred was found and this information was used when setting the NodeStatus collisions to old obstacle. This prevented the loops from repeating the limits of all three angles. This saved 1.2 seconds in interpreted Quick Basic. The NodeStatus were searched to find collisions (*ie* bit 2 set). These NodeStatus were then changed to old obstacle (*ie* all other bits were set to zero except bit one), otherwise the NodeStatus was reset to zero as shown below.

```
IF NodeStatus ( t1%, t2%, t3%) AND 2 =2 THEN
   NodeStatus ( t1%, t2%, t3%) =1
ELSE
   NodeStatus ( t1%, t2%, t3%) =0
END
```

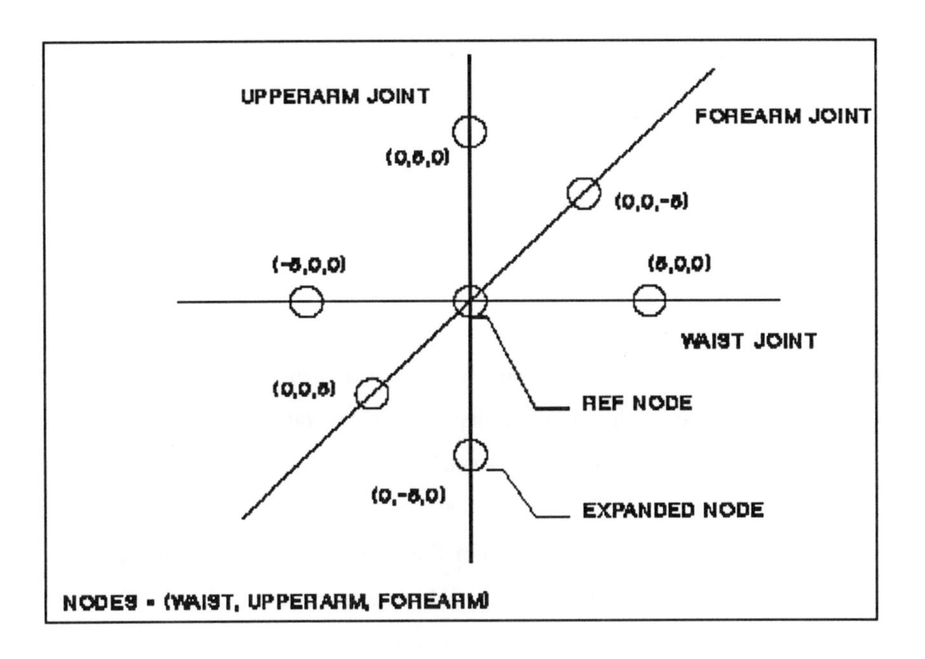

Figure **A.1**: The Expandout operation.

Appendix B

Edge Following and Line Fitting

The routines used for the limited work with stereo vision and the single camera at an angle, started at an arbitrary edge point and scanned around its immediate 8 neighbours in the plane to find a linking point. If no neighbouring point was found, the edge was said to be complete and another starting point was found. If more than one edge point was found (such as at a corner or meeting point) the routine followed one of the points whilst placing the other onto a list for future tracing. The program retained the gradient of the line it was fitting and searched its nearest neighbouring pixels for a point which continued this gradient. If this was not found, then an arbitrary pixel within the 8 was chosen whilst the pixels not selected were stored on a separate list which was expanded later. Each pixel checked was reset to a value which caused it to be undetectable to the program and thus not retraced. This technique provided an array of linked x and y edge points which was used to generate straight line information for a parallelepiped description of the object.

Line fitting used the data obtained from the edge trace routine to mathematically define vertices and their crossing points. The procedure used the 'least squares' process to match straight lines, fitting the $y=mx+c$ formula from the edge descriptions generated from the local operator in the edge detection sequence.

$$v=y-y1=ax1+b-y1$$

$$v=ax1+b-y1$$

$$v^2=(ax1+b-y1)^2$$

$$\Sigma v^2=\Sigma(ax+b-y)^2=S$$

a and b were selected so that S was zero

$$\delta S/\delta a=\Sigma 2(ax+b-y)x$$

$$\text{derivative}=0 \text{ if } a\Sigma x^2+b\Sigma x-\Sigma xy=0$$

also

$$\delta S/\delta b=\Sigma 2(ax+b-y)$$

$$\text{derivative}=0 \text{ if } a\Sigma x+bn-\Sigma y=0$$

where n is the number of points to be fitted.

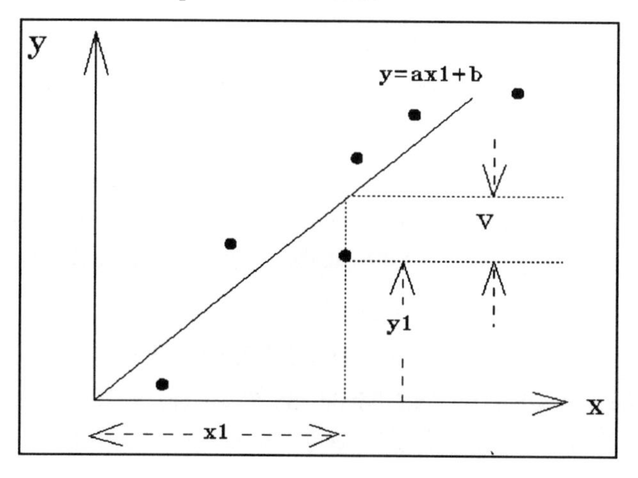

Figure B.1: The Line Fitting Process.

From these equations, a and b may be found:

$$a=\frac{n\Sigma xy-\Sigma x\Sigma y}{n\Sigma x^2-(\Sigma x)^2}$$

and

$$b=\frac{\Sigma y-r\Sigma x}{n}$$

$$\text{regression coefficient}=\frac{nxy-\Sigma x\Sigma y}{\sqrt{\{n\Sigma x^2-(\Sigma x)^2\}\{n\Sigma y^2-(\Sigma y)^2\}}}$$

Appendix C

The Mechanical Detail Of the Prototype Manipulator

This appendix includes design drawings of the prototype manipulator used during the work described in this book; the robot base with simulated arm.

Figure **C.1**: Page 249 Design drawings of the Prototype Robot Base.
Figure **C.2**: Page 250 Design drawings of the Prototype Robot Base.
Figure **C.3**: Page 251 3 Projections of the Simulated link.
Figure **C.4**: Page 252 3-D View of the Prototype Robot Base, showing the positioning of the simulated links.
Figure **C.5**: Page 253 Side view of the Prototype Robot Base, showing the positioning of the simulated links.
Figure **C.6**: Page 254 3-D View of the Prototype Robot Base, showing the positioning of the simulated links.

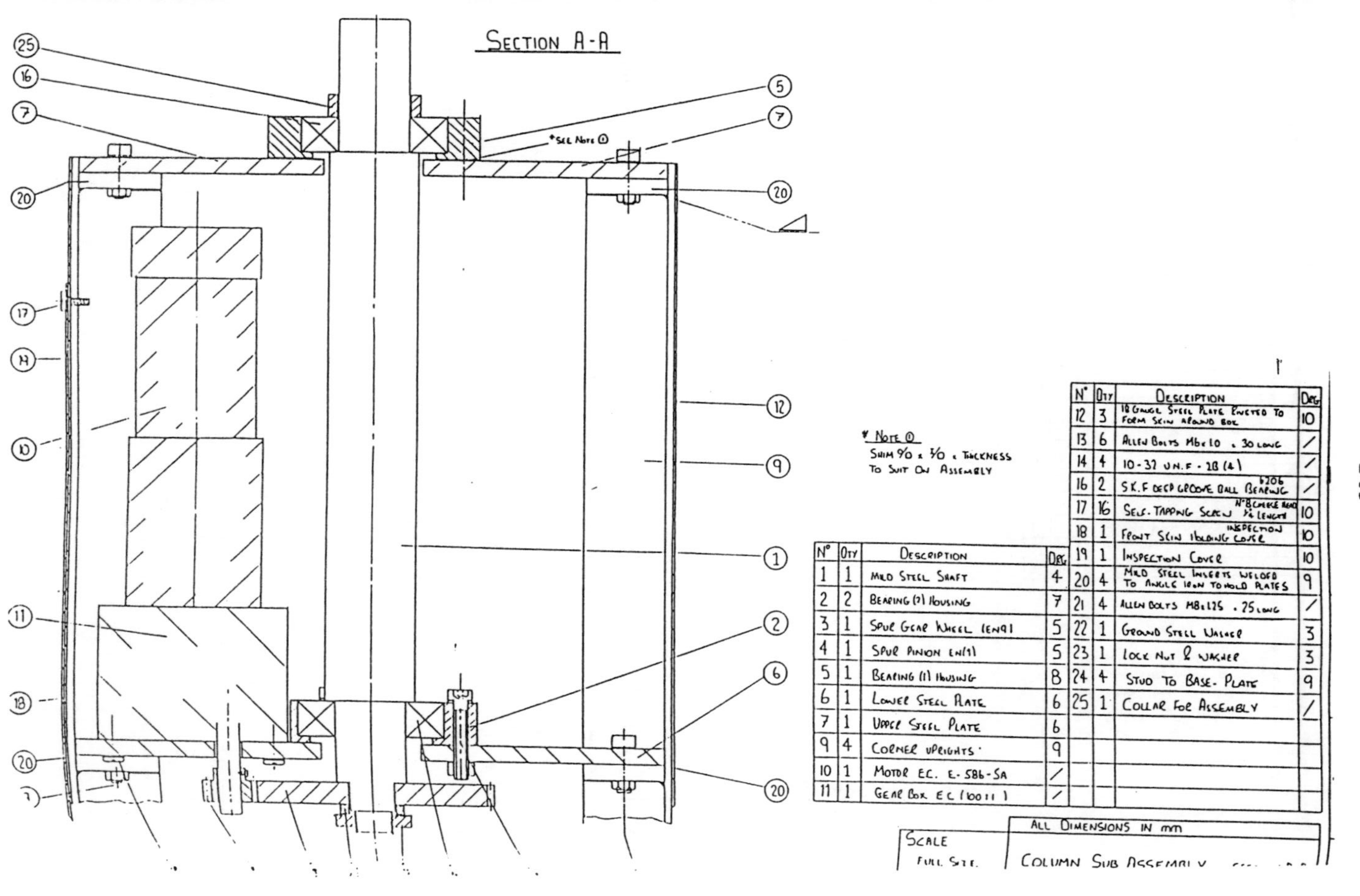

N°	QTY	DESCRIPTION	DRG
1	1	MILD STEEL SHAFT	4
2	2	BEARING (2) HOUSING	7
3	1	SPUR GEAR WHEEL (EN9)	5
4	1	SPUR PINION EN(1)	5
5	1	BEARING (1) HOUSING	8
6	1	LOWER STEEL PLATE	6
7	1	UPPER STEEL PLATE	6
9	4	CORNER UPRIGHTS	9
10	1	MOTOR EC. E-586-SA	/
11	1	GEAR BOX EC (10011)	/

N°	QTY	DESCRIPTION	DRG
12	3	18 GAUGE STEEL PLATE RIVETED TO FORM SKIN AROUND BOX	10
13	6	ALLEN BOLTS M6x1.0 x 30 LONG	/
14	4	10-32 U.N.F - 2B (4)	/
16	2	S.K.F DEEP GROOVE BALL BEARING 6206	/
17	16	SELF-TAPPING SCREW N°8 CHEESE HEAD 1/2 LENGTH	10
18	1	FRONT SKIN HOLDING INSPECTION COVER	10
19	1	INSPECTION COVER	10
20	4	MILD STEEL INSERTS WELDED TO ANGLE IRON TO HOLD PLATES	9
21	4	ALLEN BOLTS M8x1.25 x 25 LONG	/
22	1	GROUND STEEL WASHER	3
23	1	LOCK NUT & WASHER	3
24	4	STUD TO BASE-PLATE	9
25	1	COLLAR FOR ASSEMBLY	/

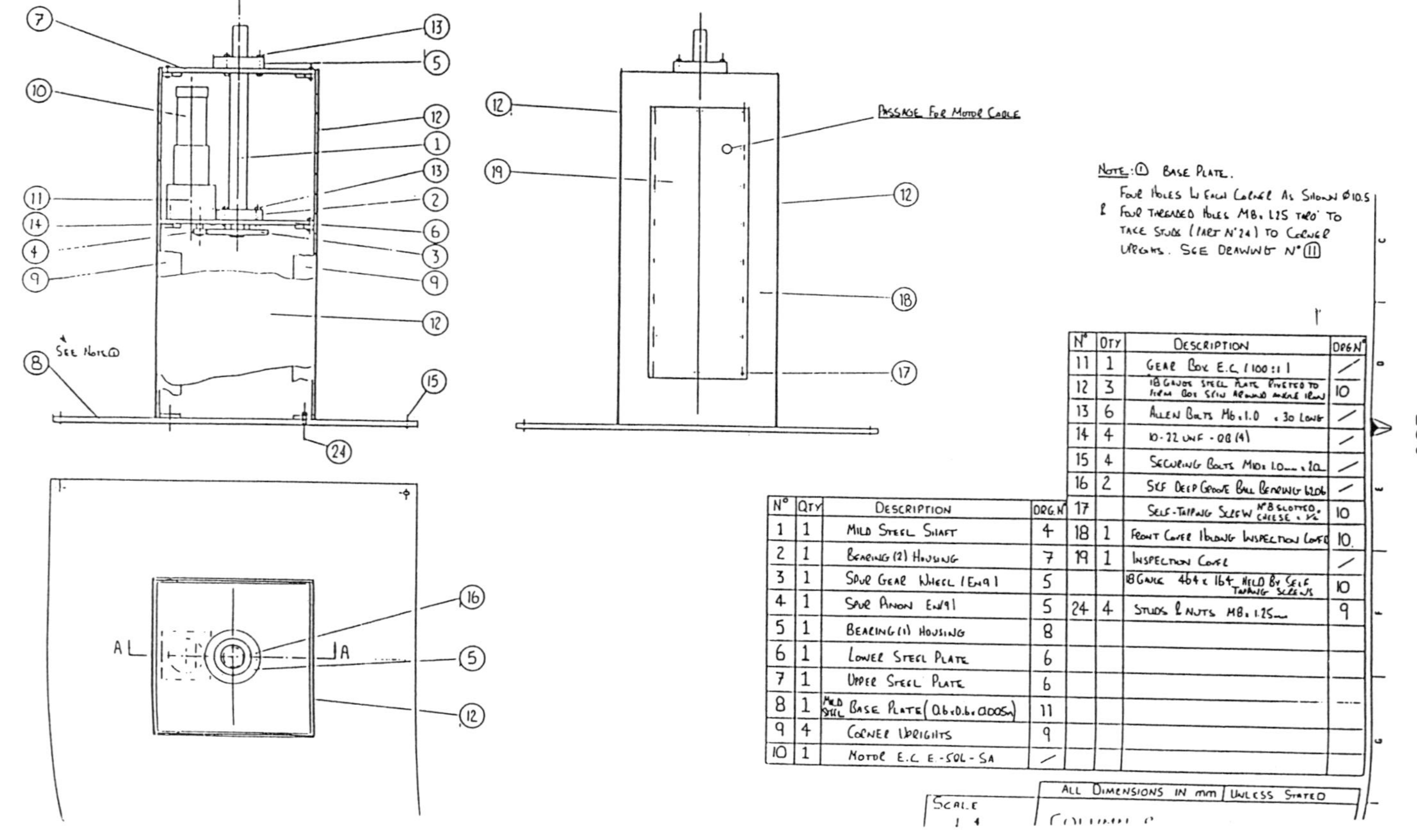

N°	Qty	Description	Drg N°
1	1	Mild Steel Shaft	4
2	1	Bearing (2) Housing	7
3	1	Spur Gear Wheel (En9)	5
4	1	Spur Pinion En9	5
5	1	Bearing (1) Housing	8
6	1	Lower Steel Plate	6
7	1	Upper Steel Plate	6
8	1	Mild Steel Base Plate (0.6 x 0.6 x 0.005m)	11
9	4	Corner Uprights	9
10	1	Motor E.C E-50L-SA	/

N°	Qty	Description	Drg N°
11	1	Gear Box E.C (100:1)	/
12	3	18 Gauge Steel Plate Riveted to Form Box Skin Around Angle Iron	10
13	6	Allen Bolts M6 x 1.0 x 30 Long	/
14	4	10-22 UNF - 0B (4)	/
15	4	Securing Bolts M10 x 1.0mm x 20	/
16	2	SKF Deep Groove Ball Bearing 6206	/
17		Self-Tapping Screw N°8 Slotted, Cheese x ½	10
18	1	Front Cover Holding Inspection Cover	10
19	1	Inspection Cover	/
		18 Gauge 464 x 164 Held By Self Tapping Screws	10
24	4	Studs & Nuts M8 x 1.25mm	9

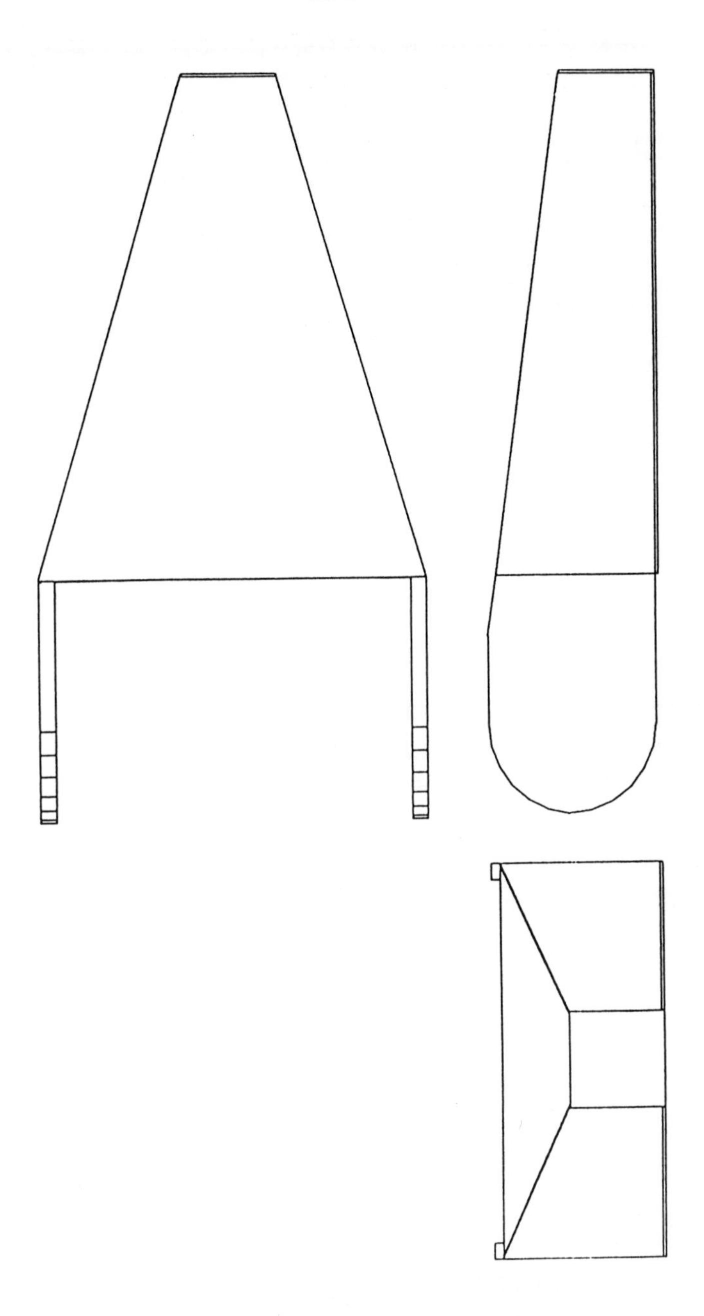

Figure C.3: 3 Projections of the Simulated link.

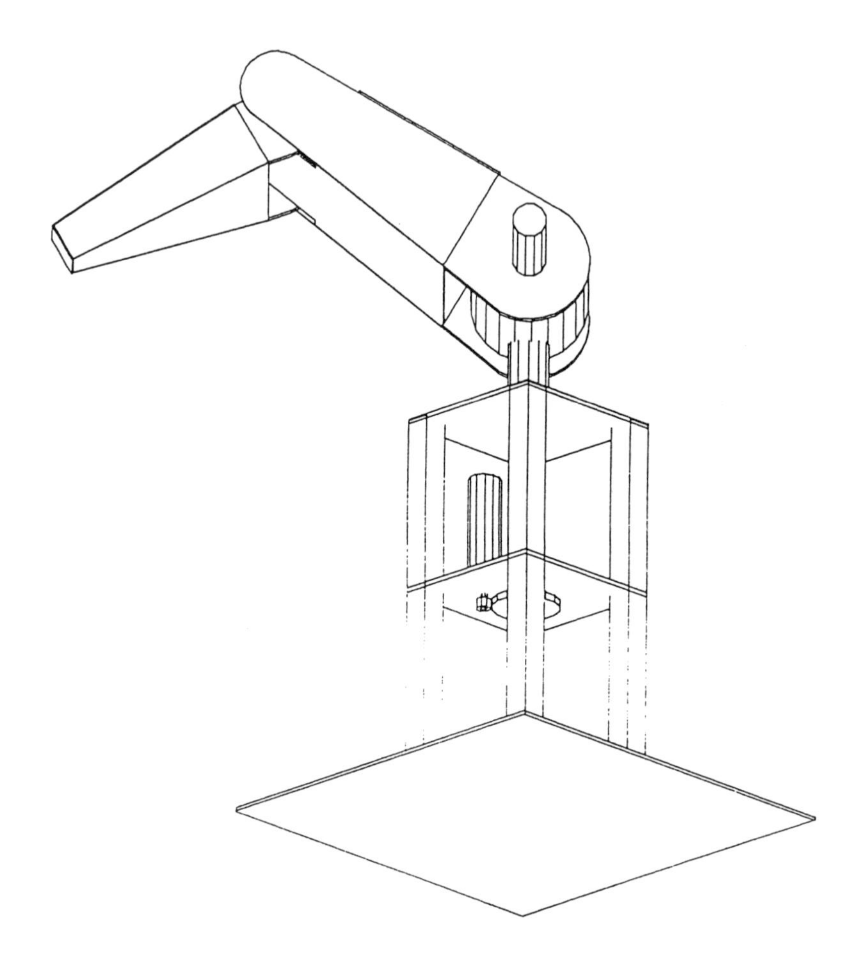

Figure C.4: 3-D View of the Prototype Robot Base, showing the positioning of the simulated links.

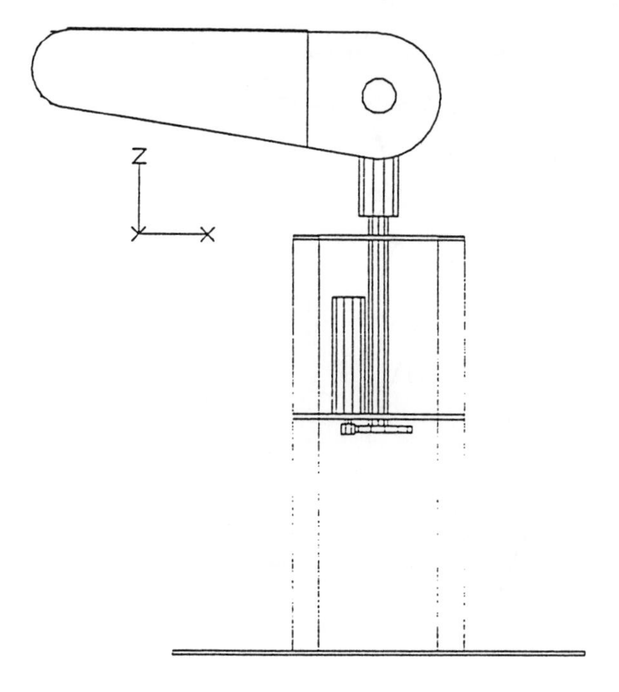

Figure C.5: Side view of the Prototype Robot Base, showing the positioning of the simulated links.

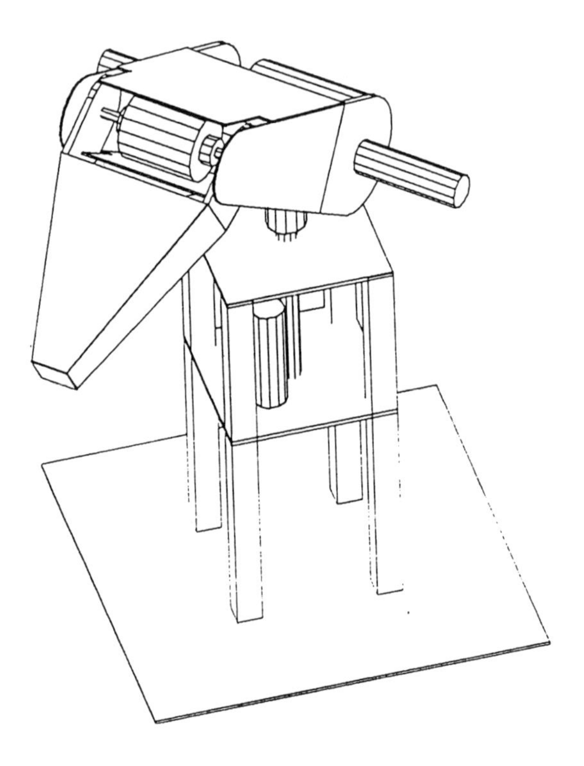

Figure C.6: 3-D View of the Prototype Robot Base, showing the positioning of the simulated links.

Appendix D

Specifications of the Mitsubishi RM-501 robot

This appendix includes the specifications of the robot used during the work described in this book; the Mitsubishi RM 501 robot.

Figure **D.1**: Page 256 The Outer Appearance of the Mitsubishi Robot.

Figure **D.2**: " The interior of the Mitsubishi Robot.

Figure **D.3**: Page 257 The Outer Dimensions and Specifications of the Mitsubishi Robot.

Figure **D.4**: Page 258 A photograph of the Mitsubishi Robot.

(1) Outer appearance

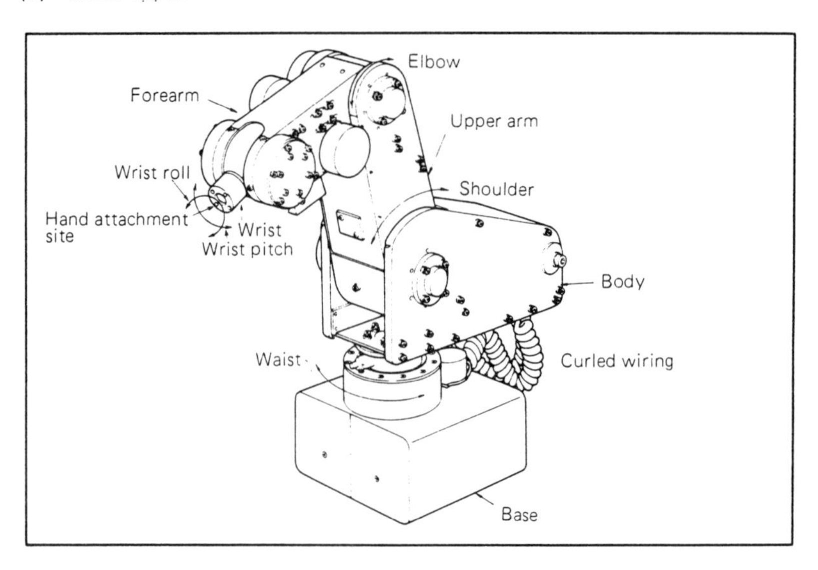

(2) Interior of arm

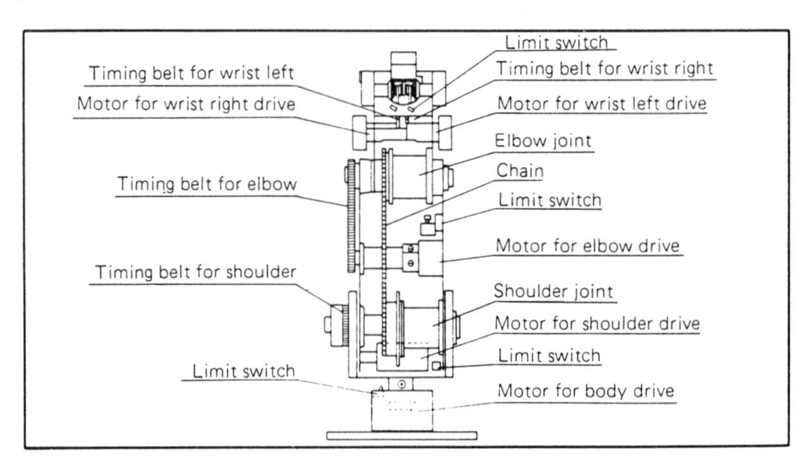

Figure **D.1**: The Outer Appearance of the Mitsubishi Robot.

Figure **D.2**: The interior of the Mitsubishi Robot.

Item		Specification
Structure		Five degrees of freedom Vertical multi-joint type
Range of movement	Waist rotation	300°
	Shoulder rotation	130°
	Elbow rotation	90°
	Wrist pitch	± 90°
	Wrist roll	± 180°
Permissible handling weight		max. 1.2 kg (includes weight of hand)
Maximum synthesis speed		400 mm/sec (wrist tool surface)
Position repeat accuracy		±0.5 mm (wrist tool surface)
Drive system		Electroservo drive by a DC servomotor
Main unit weight		about 27 kg

Note: The permissible handling weight (1.2 kg) is the value at a point 100 mm from the wrist tool surface.

Outer dimension diagram

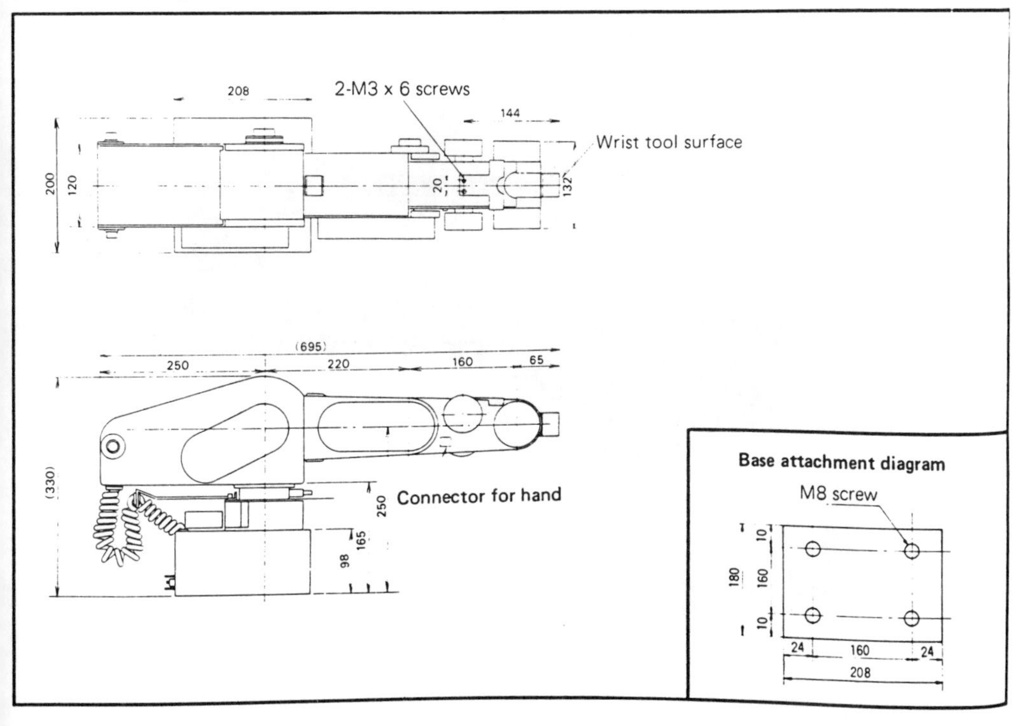

Figure D.3: The Outer Dimensions and Specifications of the Mitsubishi Robot.

Figure D.4: A photograph of the Mitsubishi Robot.

REFERENCES

ADAMS D 1979 "The Hitchhikers Guide to the Galaxy" ISBN 0-330-25864-8, Pan Books Ltd, London UK.

AHUJA et al **1980** "Interference Detection and Collision Avoidance among 3-D Objects", 1st Annual Conf on AI, Stanford Univ, pp 44-48.

AKMAN V **1985**(PhD) "Shortest Paths Avoiding Polyhedral Obstacles in 3-D Euclidean Space", Rensselaer Polytechnic Institute, New York, USA.

AN, ATKENSON and **HOLLERBACH 1986** "Estimation of Inertial Parameters of Rigid Body links of Manipulators", AI Memo 887, MIT, Feb 86.

BAJAJ & **MOH 1988** "Generalised Unfoldings for Shortest Paths", Int Journal of Robotics Research, pp 71-76.

BALDING N.W. **1987**(PhD) "Real Time Path Planning for Robot Arms", Durham (UK).

BALDING & **PREECE 1986** "Real Time Collision-Free Path Calculation", Proc of IMechE UK pp61-68.

BALL A **1983** "Designers guide to shapes", CADCAM Int, Mar 83, pp 32-34 and Oct 83 pp 42 - 44.

BEJCZY A.K **1974** "Robot Arm dynamics and Control", NASA - JPL Technical Memorandum, 33-669.

BILLINGSLEY J et al. **1983** "The Craftsman Robot", Electronics and Power, UK, December 1983.

BOBROW et al **1983** "On the optimal control of Robotic Manipulators with actuator constraints", Proc of American Control Conf pp 782-787.

BONNEY et al **1985** "Evaluation and use of a Graphical Robot Simulator - A case study from ITT/AMTC using GRASP", Int Conf on Simulation in Manufacturing, UK.

BOYSE J.W **1979** "Interference detection among solids and surfaces", Communications of the ACM, Jan 79, Vol 22, No 1. pp 3 - 9.

BRADY et al. **1979** "Planning and Execution of Straight Line Manipulator Trajectories". IBM Research Journal No 23.

BRADY et al. **1982** "Robot Motion: Planning and Control", MIT Press, ISBN 0-262-02182-x.

BRADY et al. **1989** "Robotics Science", MIT Press, ISBN 0-262-02284-2.

BRAID I.C **1973**(PhD) "Designing with Volumes", Cambridge University, UK.

BRAID I.C **1975** "The Synthesis of Solids bounded by many Faces", Communications of the ACM, Vol 18, pp 209 -221.

BROOKS R.A. **1983(a)** "Find Path for a Puma-Class Robot", Proc AAAI, Washington, USA, pp40-44.

BROOKS R.A. **1983(b)** "Solving the Find path Problem by Good Representation of Free Space", Proc of IEEE, Trans on Systems, Man and Cybernetics, Vol SMC-13, pp 190-197

BROOKS & LOZANO-PEREZ 1983(c) "A Subdivision Algorithm in Configuration Space for FindPath with rotation", Proc of Int Joint Conf AI, Karlsruhe (Germany), pp 799-806.

BROOKS R.A. **1983(d)** "Planning Collision Free Motions for Pick and Place Operations", Int Journal of Robotics Research, Vol 2, No 4, pp 19-44.

BROST R.C. **1986** "Automatic Grasp Planning in the Presence of uncertainty", IEEE Conf on Robotics and Automation, Proc of IEEE, pp 1575-1581.

BYUNG & KANG 1984 "An efficient minimum time robot path planning under realistic conditions", Proc of American Control Conf, San Diego, USA, pp 296-303.

CAMERON S. **1982** "The Clash Detection Problem", Dept of AI, University of Edinburgh, Working Paper No 126.

CANNY J. **1985** "A Voronoi Method for the Piano-Movers problem", Proc of IEEE Int Conf on Robotics and Automation, St Luis, USA, pp 530-535.

CANNY J. **1986** "Collision Detection for Moving Polyhedra", IEEE Trans on Pattern Analysis and Machine Intelligence, Vol PAMI-8, No 2.

CANNY J. **1987** "Exact Solution of Some Robot Motion Planning problems", AI Lab MIT, Computer Sci Div, Univ Cal, USA pp 505-513.

CANNY & REIF 1987 "New Lower Bound Techniques for Robot Motion Planning Problems", 28th Ann Symp on Foundations of Comp Sci, pp 49-60.

CESARONE J. **1988**(PhD) "Manipulator Collision Avoidance by Dynamic Programming", North-Western University.

CHIEN et al **1984** "Planning Collision Free Paths for Robotic Arm Among Obstacles", IEEE Trans on Pattern Analysis and Machine Intelligence, Vol PAM-6, No 1, pp 91-96.

COLLIE AA et al. **1986** "The Development of a Pneumatically Powered Walking Base", Proceedings of IMechE C337/86.

CRAIG J.J. **1989** (*Second Edition*) "Introduction to Robotics: Mechanics and Control", ISBN 0-201-09528-9, Addison Wesley, USA.

DE PENNINGTON et al **1983** "Geometric Modelling: A contribution towards intelligent robots", Proc 13th Int Symp on Industrial Robots, April 17th - 21st, Chicago. pp 7.1 - 7.15.

DONALD B.R. **1983** "Hypothesizing Channels Through Free Space in Solving the Find Path Problem", Internal Report, AI Lab, Report No AIM-736.

DONALD B.R. **1984** "Motion Planning with Six Degrees of Freedom", MIT AI Lab, Tech Report No 791.

DONALD B.R **1985** "On Motion Planning with six degrees of freedom", Proc of IEEE Int Conf on Robotics and Automation, pp 536-541.

DONALD B.R. **1987** "A Search Algorithm for Motion Planning with six degrees of Freedom", Artificial Intelligence No 31, pp 295-351

DONALD B.R. **1989** "Error Detection and Recovery in Robotics", Springer-Verlag, ISBN 3-540-96909-8.

DOTY & GOVINDARAJ 1982 "Robot obstacle detection and avoidance determined by Actuator Torques and Joint Positions", Proc of IEEE, CH1749-1/82/0000-0470, pp 470-473.

DUBOWSKY and **BLUBAUGH 1985** "Time Optimal Robotic Manipulator Motions and Work Places for point to point", Proc 24th IEEE Conf on Decision and Control, Fort Lauderdale, USA, Dec 11-13th.

DUBOWSKY, NORRIS AND **SHILLER 1986** "Time optimal trajectory planning for robot manipulators with obstacle avoidance: A CAD Approach", IEEE Int Conf on Robotics and Automation, Apr 7-10th, paper no 276.

DU PONT P.E. **1988**(PhD) "Planning Collision-Free Paths for kinematically redundant robots by selectively mapping configuration space",

DLABKA et al. **1988** "A Practical Approach for Planning and realisation of Optimal trajectories for industrial robots", Int Symp on Robot Control, pp 74.1-74.6.

FAIRHURST M.C. **1988** "Computer Vision for Robotic Systems", Prentice Hall International (UK), ISBN 0-13-166927-3.

FAVERJON B **1984** "Obstacle avoidance using an OCTREE in the configuration space of a manipulator", IEEE Conf on Robotics, Atlanta, pp 504-512.

FAVERJON B **1986** "Object Level Programming of Industrial Robots", Proc of the IEEE Int Conf on Robotics and Automation, pp 1406-1411.

FIKES et al **1972** "Learning and executing generalized robot plans", AI No 3, 251-288.

FREUND & HOYER 1984 "Collision Avoidance for Industrial Robots with Arbitrary Motion", Int Journal of Robotics Syst, No 1.4, pp 317-329.

FREUND & HOYER 1988 "Real-Time Path-Finding in Multi-robot systems including obstacle avoidance", Int Journal of Robotics Research, Vol 7, No 1, pp 42-70.

FU, GONZALEZ & LEE 1987 "Robotics: Control, Sensing, Vision and Intelligence", McGraw Hill International, ISBN 0-07-022625-3.

GALBIATI L.J **1990** "Machine Vision and Digital Image Processing Fundamentals", Prentice Hall International, ISBN 0-13-541988-3.

GILBERT & JOHNSON 1985 "Distance functions and their application to robot path planning in the presence of obstacles", IEEE Journal of Robotics and Automation, Vol RA-1, No 1, pp 21-30.

GOUZENES L. **1984** "Strategies for Solving Collision-Free Trajectories Problems for mobile and manipulator robots", Int Jnl of Robotics Research, No 3.4, pp 51-65.

GROOVER et al **1986** "Industrial Robotics", McGraw Hill Int, ISBN 0-07-024989-X.

GROSSMAN D.D. **1978** "Interactive Generation of Object Models with a manipulator", IEEE Trans on Systems, Man and Cybernetics, Vol SMC-8, No 9, pp 667-679.

HADDAD F.B **1985**(PhD) "Design and Performance of a Position Controlled Manipulator", University of Sussex, UK.

HAJASE A J **1987**(PhD) "Optimal Robot Trajectory Planning using Dynamic Models", Syracuse University, USA.

HANSEN J A et al **1983** "Generation and Evaluation of the Work-space of a Manipulator", Int Jnl of Robotics Research, No 2.3, pp 22-31.

HART, NILSSON and **RAPHAEL 1968** "A formal basis for the heuristic determination of minimum cost paths", IEEE Trans of Sys, Sci and Cybernetics, Vol 4-2, Jul 68. pp 100 - 107.

HASEGAWA T **1985** "Collision avoidance using characterized description of free space", Conf on Advanced Robotics, Tokyo, pp 69-76.

HONG Z **1986**(PhD) "Design and implementation of a Robot Force and Motion Server", Purdue University, USA.

HOPCROFT et al **1983** "Efficient detection of intersections among spheres", Int J of Robotics Research, Vol 2-4, pp 77 - 80.

HOPCROFT et al **1984**(a) "Movement problems for 2D Linkages", Soc For Ind and Applied Maths, Journal of Computing, Vol 13, No 3, pp 610-629.

HOPCROFT et al **1984**(b) "On the Complexity of Motion Planning for Multiple Independent Objects; PSPACE Hardness of the Warehouseman's Problem", Int Journal of Robotics Research, Vol 3, No 4, pp 76-88.

HOYER H. **1985** "On-Line Collision Avoidance for Industrial Robots", Proc of 1st IFAC Symp on Robot Control, pp 477-485.

HWANG K.Y **1988**(PhD) "Robot Path Planning using Potential Field Representation", University of Illinois, USA.

IZAGUIRRE & PAUL 1985 "Computation of the Inertial and Gravitational Coefficients of the Dynamics", Proc IEEE Conf on Robotics and Automation, pp 1024 - 1032.

KAMBHAMPATI & DAVIS 1986 "Multiresolution path planning for mobile robots", IEEE Jnl of Robotics and Automation, Vol RA-2, No 3, pp 135-145.

KANTABUTRA & KOSARAJU 1984 "Algorithms for robot arm movements", Proc Conf on Information Sci and Systems, Princeton University, pp 338-342.

KAHN W.E **1969**(PhD) "The near minimum time control of Open Loop Articulated Kinematic Chains" Comp Sci Dept, Stanford University, USA.

KHATIB O. **1986** "Real-Time Obstacle Avoidance for Manipulators and Mobile Robots", Proc of IEEE Int Conf on Robotics and Automation, pp 500-505. AND Int Journal of Robotic Research, Vol 5, No 1, pp 90-98.

KHATIB & **Le MAITRE 1978** "Dynamic control of manipulators operating in a complex environment", 3rd CISM IFTOMM Symp, pp 267-282.

KIM & **SHIN 1985** "Minimum Time Path Planning for Robot Arms and their Dynamics", IEEE Trans on Sys, Man and Cybernetics, Vol 15 - 2, Apr 85. pp 213 - 223.

KLAFTER,CHMIELEWSKI & **NEGIN 1989** "Robotic Engineering: An Integrated Approach", Prentice Hall International, ISBN 0-13-782053-4.

KOREN YORAM. 1985 "Robotics for Engineers" Kingsport Press ISBN 0-07-035399-9 and "Robotics for Engineers" McGRAW HILL.

KREIFELDT JG **1984** "Collision Avoidance in Robot Environments", Proc of the 1984 National Topical Meeting on Robotics and Remote Handling in Hostile Environments, pp 341-347.

KUMAR S. **1988**(PhD) "Motion Planning With Obstacles and Dynamic Constraints", Cornell University, USA.

LEU M.C **1984** "Computer Graphic Simulation of Robot Kinematics and Dynamics", Proc Robots '8' Conf, Vol 1, pp 4.80-4.101.

LEVEN D **1985** "An efficient and simple motion planning algorithm for a ladder moving in 2-D space amidst Polygonal Barriers", Proc 1st ACM Symp on Computational Geometry, pp 221-227.

LOZANO-PÉREZ & **WESLY 1979** "An Algorithm for Planning Collision-Free paths among Polyhedral Obstacles", Communications of the ACM, Vol 22, No 10, pp 560-570.

LOZANO-PÉREZ T **1981** "Automatic Planning of Manipulator Transfer Movements", IEEE Trans on Systems, Man and Cybernetics, Vol SMC-11, No10, pp 681-698.

LOZANO-PÉREZ T **1983** "Spatial Planning : A Configuration Space Approach", IEEE Trans on Computers, Vol C-32, No 2, pp 108-120.

LOZANO-PÉREZ et al **1984** "Automatic Synthesis of Fine-Motion Strategies for Robots", Int Journal of Robotics Research, Vol 3, No 1, pp 3-19.

LOZANO-PÉREZ T **1985** "Motion Planning for Simple Robot Manipulators", Proc 3rd Int Symp on Robotics Research, (France) pp 133-140.

LOZANO-PÉREZ T **1987** "A Simple Motion-Planning Algorithm for General Robot Manipulators", IEEE Journal of Robotics and Automation, Vol RA-3, No 3, pp 224-237.

LUK et al **1988** "Stochastic Force Sensing applications in Robotics", Proc of Euromicro 88, Zurich: Pub North Holland, Sep 88, pp 419 - 423.

LUMELSKY V.J. **1987** "Effect of Kinematics on Motion Planning for Planar Robot Arms Moving Amidst Unknown Obstacles", IEEE Journal of Robotics and Automation, Vol RA-3, No 3, Jun 87. pp 207 - 223.

MADDILA S.R. **1986** "Decomposition Algorithm for moving a ladder among rectangular obstacles", Proc of IEEE Int Conf on Robotics and Automation, San Fransisco, Apr 86, USA.

MAYEDA et al 1984 "A new identification method for serial Manipulator Arms", Proc of the IFAC 9th Triennial World Congress, Budapest, pp 2429 - 2434.

MAZHARSOLOOK E, **ROBINSON** DC & **SANDERS** DA. **1991** "Automatic Test, Diagnostic and Control System based on a PC, Proc of 14th IASTED Int Symp on Manufacturing and Robotics, Switzerland. June 91, pp 09-11.

MOORE E F **1959** "The shortest path through a Maze", Annals of the Computation Lab of Harvard University, No 30, pp 285-292.

MORAVEC h.p **1979** "Visual mapping by a robot rover", Proc 6th Int Joint Conf on AI (Tokyo) pp 598-600.

MUKERJEE & **BALLARD 1985** "Self-Calibration in Robotic Manipulators", Proc IEEE Conf on Robotics and Automation, St Louis, USA, pp 1050 - 1057.

NAGHDY F et al. **1985** "Craftsman Integrated Manufacture". Quality Assurance UK, Vol II June 85.

NAGHDY F et al. **1986** "Robot Force Sensing Using Stochastic Monitoring of the Actuator Torque". Robots & Automated Manufacture, IEE Control Engineering Series 28, UK.

NAGHDY F and **WU. 1987** "Interpretation of Waveforms generated by Robot Actuator Torques" Proc 1987 International Symposium on Cybernetics.

NEUMANN and **KHOSLA**. **1985** "Identification of Robot Dynamics". Proc 4th Yale Workshop on Applications of Adaptive Systems Theory, New Haven, May 85, pp42-49.

NGUYEN VAN-DUC N. **1984** "The Find-Path Problem in the Plane". A.I Memo No 760, MIT AI Lab.

O'DUNLAING SHARIR and **YAP(1984(a))** **1984** "Generalised Voroni Diagrams for Moving a Ladder I: Topological Analysis", Tech Rep 139, Courant Inst of Math Sci, New York Univ, USA.

O'DUNLAING SHARIR and **YAP(1984(b))** **1984** "Generalised Voroni Diagrams for Moving a Ladder I: Efficient Construction of the Diagram", Tech Rep 140, Courant Inst of Math Sci, New York Univ, USA.

OLSON and **BEKEY** **1985** "Identification of Parameters in Models of Robots With Rotary Joints", Proc of IEE Conf on Robotics and Automation, St Louis, pp 1045 - 1049.

ORIN and **SCHRADER** **1984** "Efficient Jacobian Determination for Robot Manipulators", Robotics Research: The 1st International Symposium, MIT Press

PAPADIMITRIOU C.H **1985** "An algorithm for shortest path motion in 3D", Information Processing Letters, No 20, pp 259-263.

PAUL R.P. **1979** "Manipulator Cartesian Path Control", IEEE Trans on Sys, Man & Cybernetics, Vol 9, No 11. pp 702 - 711.

PAUL R.P. **1981**(a) "Robot Manipulators: Mathematics, Programming and Control". MIT Press.. ISBN 0-262-16082

PAUL R.P. **1981**(b) "Differential Kinematic Control Equations for Manipulators", Trans Systems, Man Cybernetics.

PAUL et al **1983** "Dynamics of Puma Manipulator", Proc Ameriacn Control Conf 83, pp 491 - 496.

PIEPER D.L **1969**(PhD) "The Kinematics of Manipulators under Computer Control", Stanford University, (USA).

RAIBERT & **CRAIG** **1981** "Hybrid Position/Force Control of Manipulators", ASME Journal of Dynamic Systems, Measurement and Control, June 81.

RED W.E **1984** "Configuration Maps for Robot Task Planning", Computers in Engineering, pp 115-124.

REIF J.H. **1979** "Complexity of the Mover's Problem and generalisations", Proc IEEE Foundations Comp Sci (San Juan), pp 421-427.

REQUICHA A.A.G **1977** "Constructive Solid Geometry", Rochester University, NY Production Automation Project, Nov 77. NSF/RA-770476.

ROBINSON DC **1987** "Craftsman Computer Integrated Quality and Manufacture", 2nd Int Conf on Robotics and Factories of the Future, San Diego, USA, Jul 1987.

ROBINSON DC, **MAZHARSOLOOK** E & **SANDERS** DA **1991** "A computer based training (CBT). system for engineering education", Proc of the East/West Congress on Eng' Education, Kracow, Poland. Sept 91, pp 429 - 431.

SANDERS DA & **HARRISON** DJ. **1992** "The requirement for integrated and systems engineers in a complex and flexible future", Proc of the 3rd World Conference on Engineering Education, Portsmouth, UK. Sept 92.

SANDERS DA, **TEWKESBURY** G, **ROBINSON** DC and **SHERMON** D. **1992** "Managing change in engineering education within a new and competitive UK environment", Proc of the 3rd World Conference on Engineering Education, Portsmouth, UK. Sept 92.

SANDERS DA, **BEVAN** N and **WHITE** TS **1992** "A platform for sensors" Proc of the Dedicated Conference on Mechatronics (part of ISATA 92), Florence, Italy, June 92.

SANDERS DA, **HARRIS** P & **MAZHARSOLOOK** E **1992** "Image modelling in real time using spheres and simple polyhedra", IEE Proc of the 4th International Conference on Image Processing and its Applications, Maastricht, Netherlands. April 92.

SANDERS DA, **JAQUES** M & **CLOTHIER** H **1992** "Geometric modelling for real time flight simulator applications", IEE Proceedings, (Colloquium Digest C6-1991/129) on Advanced flight simulation, London, UK, April 1992, pp 4/1 - 4/6.

SANDERS DA **1992** "A new engineering school in a new and competitive environment", Febuary edition of Network, published by the Education Development Unit, Portsmouth, UK, Issue No 7, Febuary 92, pp 19 - 23.

SANDERS DA, **MAZHARSOLOOK** E & **BILLINGSLEY** J **1991** "Automatic motion planning within a manufacturing system", Proc of the 8th International Conference on Systems Engineering, Coventry, UK. ISBN 0905949-102, Sept 91, pp 916-923.

SANDERS DA **1991** "The selection of via points for manipulator paths using simple models of the machinery dynamics", Advances in manufacturing technology, volume VI, edited by D Spurgeon and O Apampa, ISBN 0.900458-46-1, pp 331 - 336.

SANDERS DA **1991** "The Use Of Simulation in Teaching Automation & Robotics", Proc of the East/West Congress on Engineering Education, Kracow, Poland. Sept 91, pp 379 - 383.

SANDERS DA **1991** "The Assessment of Lecturing Skills in Engineering Education", Proc of the East/West Congress on Engineering Education, Kracow, Poland. Sept 91, pp 141 - 144.

SANDERS DA & **TEWKESBURY** G **1991** "The introduction of the voluntary assessment of lecturing skills into the School of Systems Engineering", June edition of Network, published by the Education Development Unit, Portsmouth, UK, Issue No 6, July 91, pp 11 - 15.

SANDERS DA **1991** "Improving on the use of industrial CAD and simulation packages in teaching automation and robotics", IEE Proceedings (Colloquium Digest C6-1991/129) on Electronic CADMAT, London, UK, June 1991, pp 7/1 - 7/5.

SANDERS DA, **MOORE** A & **LUK** BL **1991** "A Joint Space Technique for Real Time Robot Path Planning", Robots in Unstructured Environments, various authors, published by the IEEE, pp 1683 - 1689, ISBN 0-7803-0078-5, IEEE Catalogue Number: 91TH376-4, Microfiche: 0-7803-0079-3.

SANDERS DA, **BILLINGSLEY** J & **ROBINSON** DC **1991** "Real Time Automatic Path Planning", Proc of the Dedicated Conference on Mechatronics (part of ISATA 91), Florence, Italy, May 91 pp 435 - 442.

SANDERS DA **1988** "DC Motor Servo", Feb edition of Practical Electronics, UK, pp 51-54.

SANDERS DA, **HERRING** D, **NAGHDY** F & **BILLINGSLEY** J **1987** "Advances in Adaptive Path Control using information from servo amplifier currents", IEE Proceedings 282 on Electrical Machines and Drives, ISBN 0 85296356 4, London, UK, Nov 87, pp 300 - 305.

SANDERS DA, **BILLINGSLEY** J & **NAGHDY** F **1987** "Adaptive Path Control to Minimise Current and Torque Transients in Motors Using Information from the Servo Amplifiers Connected to Robot Joints", Proc of CADCAM 87 Conference, Birmingham, UK. ISBN 0-948428-07-4 pp 355 - 363.

SAHAR & **HOLLERBACH** **1986** "Planning of minimum time trajectories for robot arms", Int J of Robotics Research, Vol 5, No 3. pp 90 - 100.

SCOTT. **1984** "The Robotic Revolution", Blackwell(Oxford & New York).

SHARIR & **SCHORR** **1984** "On Shortest Paths in Polyhedral Spaces", Proc 16th AMC Symp Theory of Comp, pp 144-153.

SHIN & **McKAY** **1986** "A Dynamic Programming Approach to Trajectory Planning of Robotic Manipulators", IEEE Trans on Automatic Control, Vol AC-31, No 6, pp 491-500.

SCHWARTZ & **SHARIR** **1983**(a) "On the Piano Movers Problem I: The case of a two-dimensional rigid polygonal body moving amidst polygonal barriers", Advances in Applied Mathematics, 36(3), pp 345-398.

SCHWARTZ & **SHARIR** **1983**(b) "On the Piano Movers Problem.II - General Techniques for computing topological properties of real algebraic manifolds", Advances in Applied Mathematics 4, pp 298-351.

SCHWARTZ & **SHARIR** **1983**(c) "On the Piano Movers Problem III: Coordinating the motion of several independent bodies: The special case of barriers", Int Journal of Robotics Research, Vol 2, No 3, pp 46-75.

SHARIR & **SHEFFI** **1984** "On the piano movers problem IV: Various decomposable 2-D Motion Planning Problems", Communications on Pure and Applied Maths, No 37, pp 479-493.

SCHWARTZ & SHARIR 1984 "On the Piano Movers Problem V: The case of a rod moving in 3D space amidst polyhedral obstacles", Communications in Pure and Applied Maths, No 37, pp 815-848.

SHWARTZ & YAP 1987 "Algorithmic and Geometric Aspects of Robotics: Volume One", ISBN 0-898859-554-1.

SHAMOS & HOEY 1975 "Closest Point Problems", Proc 16th IEEE Symp Foundations of Comp Sci, pp 151-162.

SINGH & LEU 1987 "Optimal trajectory generation for robotic manipulators using dynamic programming", Journal of Dynamic Systems, Measurement and Control, Vol 109, pp 88-96

SONEIRA M **1987** "Thermally Optimal and Torque-Limited Optimal Least time Paths for Robots" AT&T Bell Laboratories, Holmdel, New Jersy.

SNYDER WE. **1985** "Computer Interfacing and Control of Industrial Robots", Prentice Hall International

TAYLOR R.H. **1976** "A synthesis on Manipulation Control Programs from Task Level specifications", Report STAN-CS-76-560, Stanford Univ AI Lab, Comp Sci Dept, USA.

TEWKESBURY GE, **HARRISON** DJ, **SANDERS** DA, **BILLINGSLEY** J & **HOLLIS** JEL **1992** "A transputer based laser scanning system". Proc of the Singapore International Conference on Intelligent Control and Instrumentation, Singapore. February 92, Vol 2, pp 746 - 750.

TEWKESBURY GE, **STRICKLAND** P, **SANDERS** DA & **HOLLIS** JEL **1992** "Product orientated manufacturing", Proc of the Dedicated Conference on Mechatronics (part of ISATA 92), Florence, Italy, June 92.

THORPE C.E **1984** "Path Relaxation: Path Planning for a Mobile Robot", Proc of AAAI, Austin Texas (USA), pp 318-323.

TSENG C.S. **1987**(PhD) "Rapid Generation of Collision-Free paths for Robot Manipulators with simulation", University of Florida, (USA).

UDUPA S.M **1977**(PhD) "Collision detection and avoidance in computer controlled manipulators",

UICKER J.J **1966** "Dynamic Force Analysis of Spatial Linkages" Proc of Mechanisms Conference 1966.

VUKOBRATOVIC & KIRCANSKI 1982 "A method for optimal synthesis of manipulator robot trajectories", Trans of the ASM, Jnl of Dynamic Syst, Meas and Cont, Vol 104, pp 188-193.

WARREN et al **1989** "An Approach to Manipulator Path planning", Int Journal of Robotics Research, Vol 8, No(5), pp 87-95.

INDEX

A

A.I.	6, 240
ABIDI	116
ADAMS	VIII, 259
ADAPTER	9, 10, 19, 134, 162, 194
ADAPTION	110, 111, 120, 126, 224, 236, 237
AHUJA	22, 27, 259
AMPLIFIER	42, 43
APPROACH PATH	168, 169
APPROACH PLANNING	165, 166
ATKENSON	196, 259
AUTOMATIC MOTION PLANNING	4, 8, 11, 137
AUTOMATIC PATH PLANNING	8, 10, 25, 26, 133, 230
AUTOMATIC PROGRAMMING	4, 6, 190, 195, 226, 238

B

BAJAJ	18, 259
BALDING	8, 24, 26, 27, 69, 101, 259
BALL	71
BALLARD	196
BAYES THEOREM	112
BEKEY	196
BEZIER EQUATIONS	71
BILLINGSLEY	240, 259
BLOCKED CONFIGURATIONS	74, 76, 80, 90-99, 104, 105, 118, 164, 175, 179, 186-189, 231, 233
BOBROW	35, 259
BONNEY	38, 259
BOYSE	22, 259

BRADY 7, 196, 203, 240, 259
BRAID 70, 259
BROOKS 17, 20, 27, 29, 30, 74, 241, 259
BUS SYSTEM 57/58

C
CAD 6, 133, 190, 194, 195
CAMERA 13, 109, 114, 119, 135, 137, 138, 141, 143, 144, 228, 234
CANNY 27, 31, 260
CERTAINTY GRID 109, 116, 119, 231, 233
CHIEN 31, 260
CHMIELEWSKI 11/12
CLEAR CONFIGURATIONS 75-77, 81, 178/179, 186/187, 231
COLLISION 28, 129, 130
COLLISION AVOIDANCE 3-5, 8, 164, 178
COLLISION DETECTION 14, 22, 25, 130, 131, 131, 178, 236
COMMUNICATIONS 56, 185, 229, 241
CONES 17, 30, 174
CONFIGURATION SPACE 15/16, 28, 30, 76, 177
CONFIGURATION SPACE GRAPH 8, 178
CONFIGURATIONS 5, 8, 15, 76, 90-98, 164, 166, 177, 193/194
CONFLICTS 10, 14, 109, 118, 119, 233, 234
CONSTRAINTS 7, 9, 16, 134
CONSTRAINTS DYNAMIC 9
CONSTRAINTS ROBOT 71
CONSTRUCTIVE SOLID GEOMETRY 22, 70/71
CONTROL COMPUTER 9, 41, 48, 185
CONTROL SYSTEM 41, 239
CONTROLLER 5/6, 15, 41, 49, 51, 55, 66, 120/121, 129-131, 137, 165, 168, 185, 230
CORNER POINTS 169
CORRIDORS 32
COSINE RULE 78
COST FUNCTION 7, 33, 36, 171, 172, 175, 180
CRAIG 13, 236, 260
CROSSOVER DISTORTION 42
CROWLY 112
CSPACE 16, 30
CUBE 74, 97, 101/102, 187, 189
CURRENT 10, 18, 42, 52, 120-130, 195, 205, 206, 227, 235, 236
CYLINDERS 24, 72-74, 87, 92, 96, 163, 187, 189

D
DATA FUSION 111
DATA PROCESSOR 136
DAVIES 13
DE PENNINGTON 20, 22, 25, 27, 260
DECISION GRAPH 119
DETECTION 11, 12, 14, 129
DETECTION OBSTACLES 137, 138, 155
DETECTION PEAK 13
DETECTION VISION 13
DIFFERENTIAL EQUATIONS 18
DIRECTIVITY ERRORS 12, 118
DISTANCE FUNCTION 23
DISTANCE TRAVELLED 34, 171, 172
DIVERSE 226
DONALD 27, 29, 260
DOTY 11, 13, 260
DUBOWSKY 260
DUPONT 22, 27, 182, 260
DYNAMICS ENVIRONMENT 12, 14
DYNAMICS EQUATIONS 19, 206
DYNAMICS EXPERIMENTS 5, 9/10, 16, 18, 194, 196-206, 224, 226, 238
DYNAMICS OBSTACLES 11, 14, 73, 106, 132, 135

E
EDGE DETECTION 114/115, 138, 151, 229
EDGE FOLLOWING 251
EFFICIENCY 7
ELFES 116
ENVIRONMENT 9, 12, 14, 21, 116
ENVIRONMENT DYNAMIC 10-12, 14
ENVIRONMENT FUTURE 239
ENVIRONMENT STATIC 9, 11, 14, 70, 135

F
FAIRHURST 13, 265
FAVERJON 17, 260/261
FILTERING 52
FLAGS 77, 179
FLEXIBLE MANUFACTURE VI, 2, 3, 7
FORCE DETECTION 12, 120, 133
FORCE FEEDBACK 11/12
FORCE SENSING 14, 109, 129, 130, 234, 235
FORCE VECTOR 19

FREE SPACE 17, 22, 29, 69, 187
FREEWAYS 17, 30
FRICTION 13, 206, 221, 222, 224, 225
FU 196, 261
FUSION (SENSOR) 111-113, 116
FUTURE WORK 11, 238

G

G64 BUS 57/58, 229
GALBIATI 11, 13, 226
GEOMETRIC CONSTRAINTS 7
GEOMETRIC FUSION 112
GEOMETRIC MODELLING VII, 21, 24, 107, 241
GEOMETRIC PROBLEM 6
GLOBAL PATH PLANNING 9/10, 16, 29, 168, 172, 176, 178, 188, 230
GOAL CONFIGURATION 18, 26, 28, 35, 76, 164, 168/169, 172, 175, 177-185
GONZALEZ 11, 196, 261
GOTO 111
GOVINDARAJ 11, 13
GRANDJEAN 112
GRAPH 2-SPACE 8, 171/172, 175
GRAPH 3-SPACE 9, 75, 179, 180, 193
GRAPH C-SPACE 8/9, 76/77, 100, 171, 175, 178
GRAPH DECISION 119
GRASP 21, 60, 70, 190
GRAVITY 18, 207, 222
GRID CERTAINTY 116, 119, 233
GROWING TECHNIQUES 72

H

HADDAD 196, 261
HAGER 112
HARDWARE 40, 50, 166
HART 30, 169, 261
HAYNES 240
HEURISTIC PLANNING METHODS 8, 10, 26, 28, 106, 168, 177/178, 188, 190, 193, 230
HIGH LEVEL VISION TECHNIQUES 151, 229
HOLLERBACH 36, 196
HOLLIS 240
HOLLOW SPHERES 83, 85, 87
HOPCROFT 6, 73, 261
HWANG 27, 261

I

IMAGE DATA 147
IMAGE PROCESSING 41, 135, 228
IMPROVEMENT 18, 41, 120, 195, 219, 226, 235-237
INDEPENDENT FUSION 111
INDUCTIVE SENSORS 118
INDUSTRIAL SIMULATION 10
INERTIA 13, 197-203, 225
INERTIA MATRIX 18
INTERFERENCE DETECTION 21, 22, 25
INTERSECTION CALCULATION 75, 104
INVERSE KINEMATICS 106
IRREGULAR STOP 120, 124
IRRELEVANT STOP 120, 123, 128
IZAGUIRRE 196, 261

J

JACOBIAN 15, 236
JAQUES 241
JOINT ANGLES 15
JOINT MOTOR CURRENTS 10
JOINT MOVEMENT 8
JOINT SPACE 15, 17, 75, 76, 77, 81, 100, 105, 136

K

KALMAN FILTER 112
KHAN 196, 261
KHATIB 8, 23, 27/28, 231, 261/262
KHOSLA 196
KIM 35, 262
KINEMATIC CHAIN 8, 25/26, 30
KINEMATICS INVERSE 33, 164, 189/190
KINEMATICS 33, 164, 189/190
KLAFTER 11/12, 262
KRIEGMAN 111
KUAN 17
KUMAR 3, 7, 196, 262

L

LAGRANGIAN 196, 197-203
LAPLACIAN 115
LASER 114, 117, 119, 234
LEE 11, 196
LEVELS FUNCTIONS 5

LEVELS HIGH 23, 49, 10
LEVELS IMAGE PROCESSING 135
LEVELS JOINT 16
LEVELS LOW 49
LEVELS STRATEGIC 41, 49, 51, 130, 136
LEVELS SUPERVISORY 41, 49, 51, 136
LEVELS TASK 15
LIGHTING 137
LINE FITTING 143, 251
LO 139
LOCAL PATH PLANNING 8, 10, 26, 108, 177/178, 188, 190, 193, 230
LOW LEVEL VISION 145, 228
LOZANO-PEREZ 16-18, 20/21, 27, 29, 32, 262
LUH 31, 112
LUK 13, 120, 262
LUMELSKY 28, 262
LUO 117

M

MADDILA 32, 262
MAIN COMPUTER 41, 54, 126, 131, 136, 186
MCKERROW 11/12, 239
MEDIAN FILTERING 114, 146, 148/149, 232
MINIMUM DISTANCE 171
MINIMUM ENERGY 18, 33, 36
MINIMUM TIME 18, 33, 35, 110
MINTZ 112
MITSUBISHI ROBOT 48, 59-64, 72, 134, 164, 167, 177/178, 224/225, 229, 260-263
MIXER STAGE 43/44, 129
MOBILE ROBOTS 28, 30, 112, 236
MODELLING 69, 113, 194
MODELLING GEOMETRIC VII, 21, 24, 227, 241
MODELLING KINEMATIC 21
MODELLING MACHINERY 20
MODELLING POLYHEDRAL 10, 70, 193
MODELLING ROBOT 24, 70/71, 103, 190, 198, 206
MODELLING SOLID 22, 177, 193
MODELLING SURFACE 70/71
MODELLING WORLD VIII
MODELS 2-d SLICES 87, 89, 107, 187, 189, 193, 227/228
MODELS DYNAMIC 10, 69, 224, 226
MODELS OBSTACLES 21, 169
MODELS SPHERE 87

MODELS STATIC 21, 226/227
MOH 18
MOTION PLANNING 4, 7, 9, 11, 30, 41, 164/165
MOTION IMPROVEMENT 120, 132, 133, 235
MOTION GRAPH 32
MOTION TESTS 206, 208-219, 224
MOTION PLANNING AUTOMATIC 8/9, 11, 20, 162, 164/165, 226
MOTION PLANNING WHY? VI, 1
MOTOR CURRENT 10
MUKERJEE 196, 262

N
NAGHDY 120, 262
NAKAMURA 112
NAVIGATION 111
NEGIN 11/12
NEIGHBOURHOOD AVERAGING 114, 146, 148
NEUMANN 196, 262
NGUYEN 28, 262
NODES 8, 169, 172/173, 175, 177, 179/180, 186, 231
NOISE 135, 138, 234
NON-VITAL MOVE 110, 120, 128

O
OATEN 141
O'DUNLAING 31, 263
OBSTACLE AVOIDANCE 4, 11, 23, 28, 155, 168
OBSTACLE DETECTION 10-12, 20, 129, 137, 138, 145, 151, 155, 235
OBSTACLES 9, 12/13, 15-17, 73, 76, 98, 113, 138, 169, 177, 187/188
OBSTACLES DYNAMIC 14, 73, 106, 135
OFF LINE PLANNER 8
OFF LINE PROGRAMMING 2, 5, 7, 10, 16
OLSEN 196, 263
OPTICAL PROXIMITY SWITCH 117
OPTIMISATION 3, 5, 7, 33/34
ORIN 129, 263
OVERSEER 53-55, 126, 130, 136, 234

P
PAPADEMETRIOU 18, 263
PARALLELEPIPED 73/74, 83, 86, 108, 168-170, 177, 193, 228
PATH ADAPTION VII, 3, 9/10, 19, 41, 53-55, 126, 136/137, 162, 226, 236

PATH FEASIBILITY	165/166
PATH FINDING	8, 174
PATH IMPROVEMENT	7, 14, 17, 18, 41, 194-225, 236
PATH OPTIMISATION	3, 5, 33
PATH PLANNER	45, 55, 133
PATH PLANNING	VII, 5, 10, 15-18, 25, 29, 31, 41, 108, 133, 137, 164-193, 225, 229
PATH SEARCHING	8
PATH SHAPE	37, 38
PATH SHORTEST	8, 17
PATH SUB-OPTIMAL	8/9
PATTERN RECOGNITION	136, 144, 154
PAUL	36, 196, 203, 263
PEAK DETECTOR	41, 53, 55, 110, 126, 130, 136, 235
PERCEPTION	113
PERCEPTION	41, 53, 55, 110, 126, 130, 136, 235
PIEPER	20, 24, 26/27, 263
PLANNING GLOBAL	9, 29, 75
PLANNING LOCAL	26, 168
PLANNING OFF LINE	8
PLANNING PATH	VII, 5, 10, 15-18, 25, 29, 31, 41, 108, 133, 137, 164-193, 225, 229
PLANNING TIME	7
PLANNING TRAJECTORY	6, 7
POLYHEDRA	10, 16, 17, 20, 21, 30, 70, 74, 83, 86, 93, 94, 95, 142, 159, 241
POST-PROCESSING	194
PRIMARY MAP	16
PRIMITIVES	22
PROGRAMMING AUTOMATIC	7, 20
PROGRAMMING LEAD	2
PROGRAMMING OFF LINE	2, 5, 7
PROXIMITY DETECTION	12

R

RAIBERT	13, 258
RANGING	11/12, 14, 119, 141, 234
REAL TIME	7, 17, 19, 188, 190
RED	16, 263
REQUICHA	70, 263
RESULTS RULES	222
RESULTS IMAGES	158
RESULTS MODELLING	37
RESULTS MOTION TESTS	217-219, 220-222

RESULTS STATIC TESTS 210-216, 219/220
ROBOT COST 1/2
ROBOT DEFINITION VIII
ROBOT MITSUBISHI 59, 134, 164, 167, 178, 206, 221/222, 224/225, 229, 255-258
ROBOT MODEL 25, 71/72, 165, 167, 206, 221
ROBOT PROTOTYPE 172
ROBOT TRAJECTORY 3, 237
ROTATION MAPPING GRAPH 31
RULES 10, 19, 119, 219, 222-225

S
SAFE DISTANCE 12
SAFETY 18, 33, 38
SAHAR 36, 268
SANDERS 13, 19, 120, 242-246
SCHEMATIC MODEL 24
SCHORR 18
SCHRAPER 129
SCHWARTZ 6/7, 27, 30
SCOTT 236, 263
SENSORS 6, 9, 13, 109, 111
SENSORS FORCE 109
SENSORS FUSION 109, 111-113, 116, 118, 231-233
SENSORS PROXIMITY 14, 112, 117
SENSORS RANGING 14
SENSORS TOUCH 29, 112
SENSORS VISION 13
SERVO AMPLIFIER 19, 42/43
SHARIR 6, 18, 27, 30, 263
SHIN 35, 263
SIMULATION 10, 164, 169, 172, 178, 190-192
SLICES 2-D 72-74, 81, 87, 89, 107, 227/228
SMOOTHING 114, 146-148, 150, 232
SNYDER 195, 264
SOFTWARE 56, 119, 121, 131, 225, 235
SOFTWARE SYSTEMS 49
SOLID MODELLING 22, 177, 193
SPACE 2-SPACE 8, 167/168, 176-178, 193, 230
SPACE 3-SPACE 10, 29, 164, 177-186, 193, 230
SPATIAL OCCUPANCY 22
SPATIAL DOMAIN TECHNIQUES 145
SPHERES 20, 24, 72/73, 77, 87, 101, 103, 107/108, 168, 172/173, 177, 227

START CONFIGURATION 28,35,76,164/165,168-172,175,177,179,183/184
STATEGIC LEVEL 49, 51, 55, 130, 136
STATIC ENVIRONMENT 14, 70, 135
STATIC FRICTION 207
STATIC MODEL 14
STATIC OBSTACLES 11
STATIC TESTS 206/207, 224
STEREO VISION 135, 138, 144, 251
STRICKLAND 240
STRING PULLING TECHNIQUE 182, 184
SUPERVISORY LEVEL 49, 51, 55, 136
SURFACE MODELLING 70/71
SYSTEM 53, 229
SYSTEM CHARACTERISTICS 3

T
TASK 2, 4-6, 17, 55, 165, 168, 188
TASK DESCRIPTION 21
TASK LEVEL 15
TASK PLANNER 6
TAYLOR 15, 241, 264
TEMPLATE MATCHING 139
TEWKESBURY 240
THRESHOLDING 146, 150, 232
TORQUES 12-14, 18, 23, 110, 130, 133, 195, 205-207, 227, 235
TRAJECTORY 3/4, 12, 20, 131, 134, 188, 194, 236
TRAJECTORY CONTROL 13
TRAJECTORY GENERATION 185/186, 231
TRAJECTORY LOCUS 5, 25, 53, 165, 168, 178, 186
TRAJECTORY PLANNER 6/7
TRAJECTORY PLANNING 5, 7
TRANSFORMATION 15-17, 30, 75, 77, 81, 136, 177, 247
TRANSFORMATION ALGORITHMS 77/78, 100
TRANSIENTS 121-126, 130
TREE STRUCTURE 21/22, 30
TSENG 27, 29, 264

U
UDUPA 3, 11, 16, 20, 24-28, 72, 101, 165, 264
UICKER 196, 264
ULTRASONIC SWITCH 114, 117
UNNECESSARY MOVE 128, 133
UPPER ARM 16/17, 28, 31, 72, 78, 100, 177

V

VIA POINTS 3, 121, 126, 132, 185, 194/195, 224/225
VISIBILITY GRAPH 30, 32
VISION ERRORS 118
VISION FEEDBACK 13, 251
VISION SYSTEM 10/11, 13, 41, 118, 135/136, 228
VITAL MOVEMENT 110, 120, 128
VORONI DIAGRAMS 31

W

WEAR 38
WELDING 15
WESLY 16, 29, 32
WIDDOES 20
WIRE FRAME MODEL 177, 193
WORK SPACE 11-14, 75, 164
WORK CELLS 6, 164
WORK PIECE 39, 72
WORK PLACE 14, 194
WORLD MODEL 6, 9, 12, 14, 21, 24/25, 55

X

X_n 112

Y

YAP 6/7, 263

Z

ZHENG 112